# Mysteries of Electromagnetic Energy: Definitively Solved and Simply Explained

## Part 1:

## Fundamental Properties of Electromagnetic Energy

by Mark Fennell
© 2014

# Websites of Mark Fennell

1. Energy Technologies Explained Simply
    http://EnergyTechnologySimply.blogspot.com

2. Mark Fennell: Authentic Expressions of a Happy Dancer and Multidimensional Intellectual
    http://markfennell.blogspot.com

3. You Tube Channel: All Things Energy
    http://www.youtube.com/channel/UCk5ckPqF4oD0JoJMSBi2Zcg

# Energy Books by Mark Fennell

Renewable Energy Books
- Introduction to Electrical Power

- Wind Power Technology Explained Simply
- Hydropower Explained Simply
- Solar Power Technologies Explained Simply
- Practical Considerations of Solar Power
- Advanced Solar Cell Technologies

Natural Gas Books
- Natural Gas Basics
- Extracting and Refining Natural Gas (includes Fracking)
- Transportation, Storage, and Use of Natural Gas

Coal Power Books
- Formation and Mining of Coal
- Clean Coal Technologies
- Mercury and Coal Power

Nuclear Power Books
- Nuclear Power Meltdowns and Explosions
- Health Hazards of Radioactive Decay
- Radiation Measurements
- Processes of Radioactive Decay and Storage of Nuclear Waste

Power Line and Grid Books
- Introduction to the Transmission of Electrical Power
- Power Lines
- Underground Cables
- Utility Operations and Quality Control
- Power Grids Explained Simply

## © Copyright Notice ©

This work is copyrighted to the author. No part of this book can be published or presented to the public as the work of any other author.

Sections of this book *may be* reproduced for educational purposes. This includes use in courses, journals, websites, and other on-line educational material.

However, the author must be given full credit, with author's name and the title of this book, such as "Model and illustrations created by Mark Fennell, as published in *Mysteries of Electromagnetic Energy: Definitively Solved and Simply Explained.*"

When possible, a direct link to the book on Amazon will be provided.

For all other uses, including uses which create a profit or increase the public name for the individual, this author must be contacted: markpoet@aol.com

# Preface
## for
## *Mysteries of Electromagnetic Energy, Part 1: Fundamental Properties of Electromagnetic Energy*

Introduction

This book is the first in a series on major discoveries in the arena of electromagnetic energy.

For the past three years I have made numerous discoveries in the fundamental properties and processes of electromagnetic energy and electrons. These discoveries will advance our understanding of the universe in major ways.

For example, one on my first discoveries was the cause of particle-wave duality. For over a century scientists have known of the particle-wave duality of electromagnetic energy. Yet they did not know the reason. For the first time, we have the reason!

I first discovered this for the photon, then for the electron, and then created my General Principle of Particle Wave Duality.

These discoveries and many more are presented throughout the series of books. This first book describes and illustrates the Fundamental Properties of Electromagnetic Energy.

Important Discoveries in this Book

Some of the most important discoveries and properties discussed in this book include the following:

- Particle-Wave Duality: full explanation and illustrations of the solution, presented here for the first time.

    For over a century, the most brilliant minds (including Einstein and Bohr) have been puzzling over how electromagnetic energy can be both a particle and a wave. For the first time, the cause of this duality has been solved, and fully illustrated.

- The "Hidden Variables" of electromagnetic energy are finally discovered, explained, and illustrated.

    Understanding these "hidden variables" will simplify all theorems related to Electromagnetic Energy, Quantum Science, the Atom, and much more.

- The true nature of the electric and magnetic fields, fully illustrated.
    What exactly is the electric field? What exactly is the magnetic field? This has been a mystery for many years. I have discovered the true nature of these fields.

- The composition of the photon:
    What does the photon look like, if you held it in your hand?
    The answer is here, fully discussed and illustrated.

- The pulsation of electromagnetic energy:
    This is a key process in the nature of electromagnetic energy, and is demonstrated for the first time here.

- The creation of the wave pattern:
    The "wave" is one half of the dual nature of electromagnetic energy. The wave is also a central component in most of quantum mechanics. Therefore, in this section we explain and illustrate exactly how a wave, of any type, is created.

- Reassessment of the relationships between energy, frequency, and wavelength.
    The relations have long been known, but from a certain perspective. Once you understand how waves are created, and the cause for particle-wave duality, it is time to reassess the relationships among energy, frequency, and wavelength.

- Detailed discussion and illustrations of Energy Strings:
    The energy strings are very important in understanding everything related to electromagnetic energy and electrons. In fact, they are the "hidden variables" in quantum science and electromagnetic energy which has long been sought after.

    Therefore this book will spend significant time explaining and illustrating the nature and properties of these energy strings.

- Accurate understanding of "quanta" of light, including illustrations of the group spread.

- Description and illustration of all motions of electromagnetic energy

## Diagrams, Analogies, and Simple Explanations

One of the best features of this book, as well as every book in the series, is the numerous illustrations. Every concept is not merely explained in words, but also illustrated with detailed diagrams.

On average, a chapter has approximately 25 full color illustrations. Most of these illustrations are highly detailed, with numerous elements in each figure.

## Actual Objects and Actual Processes

This brings us to another tremendous asset for any reader or student: my strong preference wanting to know the physical reality.

In contrast to most physicists, who prefer to communicate their ideas through complicated mathematics, I much prefer to talk of real objects and real processes.

I am very much akin to Bohr, wanting to know the physical reality, not just the mathematics. And in the same way that Bohr developed a series of physical models of the atom, I have developed a series of physical models for electromagnetic energy.

Thus, my books have the related assets of 1) focusing on real objects (rather than complex math or abstract ideas), and 2) showing these discoveries through numerous illustrations (rather than a set of equations).

I very much hope that these illustrations, along with the explanations, will help any reader and any student to easily understand each process and property I have discovered.

## As an Asset for the Scientific Community

Therefore, I am pleased to offer these discoveries for the benefit of the scientific community and the general public. I hope that this book (as well as the other books in the series) will be easily understood.

It is my wish that these discoveries, models, and explanations will become commonplace, among both the layman readers and the scientific community, where more people can have a richer understanding of our universe.

# Series Summary
## for
## *Mysteries of Electromagnetic Energy: Definitively Solved and Simply Explained*

Introduction

For the past three years I have made numerous discoveries in the arena of electromagnetic energy and electrons. These discoveries will dramatically change the way we understand processes in the universe.

List of Major Discoveries Explained in this Series

Some of the more fundamental discoveries include:

- The true, long sought after, "hidden variables" of quantum mechanics.
    Knowledge and understanding of these hidden variables will simplify quantum mechanics and electrodynamics to an incredible degree.

- What is the exact composition of the photon?
    If you could hold a photon in your hands and study it, this is what you would see.

- The root cause for particle-wave duality:
    - for the Photon
    - and for the Electron

    In this series of books, the elusive solution for particle-wave duality is finally explained and illustrated.

- Why electromagnetic energy has a frequency.
    - Why it has a particular frequency
    - And why this frequency seems to last forever

- The exact process of emission of electromagnetic energy
    - And why one frequency will be emitted rather than another

- The exact process of absorption of electromagnetic energy

- Why an electron has internal energy
    - What causes the motions of electrons
    - What creates the electron spin

- The true nature of the electric field and the magnetic field

- More accurate understanding of electrons traveling in orbitals
    - Including a new models the molecular bond

- A more accurate understanding of traditional quantum terms such as "standing wave", "wave packets", and "continuous wave".
    These traditional terms can be understood when compared to the actual physical reality of electromagnetic energy.

These discoveries and many more will be presented to the public in this series of books. For the first time, the true physical nature of electromagnetic energy can be entirely understood.

<u>The Series: Mysteries of Electromagnetic Energy</u>
Collectively titled *Mysteries of Electromagnetic Energy: Definitively Solved and Simply Explained*, these books will explain and show numerous processes and properties related to electromagnetic energy and electrons.

This series of books has numerous diagrams, unique analogies, and simple explanations, all of which will help you understand the true nature of electromagnetic energy.

Most of the content presented in these books are discoveries which are new to the world, and I am very pleased to offer these discoveries to the public.

# List of Books for the Series
*Mysteries of Electromagnetic Energy:
Definitively Solved and Simply Explained*

Part 1: Fundamentals of Electromagnetic Energy

Part 2: Creation and Emission of Electromagnetic Energy

Part 3: Frequency and Pulsation of Electromagnetic Energy

Part 4: Penetration and Absorption of Electromagnetic Energy

Part 5: Communication Using Electromagnetic Energy

Part 6: Diffraction, Interference, and Red Shift

Part 7: Particle Wave Duality

Part 8: New Models of Electrons, Orbitals, and Atoms

# Author's Personal Introduction to the Reader

I have been interested in electromagnetic energy most of my adult life. It is an amazing entity, with numerous properties and the ability to perform a variety of interesting phenomenon.

Furthermore, I have come to realize over the years how important electromagnetic energy is to our lives. Electromagnetic energy is used both in the natural world and in man-made devices (and in thousands of ways). Electromagnetic energy is used to produce light, to create color, and to see into the most remote locations. It is used for art and entertainment, for health and medicine, and for communication. Indeed, I have come to realize that electromagnetic energy has a profound impact on most of our personal experiences.

However, it is only recently that I truly understand the nature of electromagnetic energy. This deep understanding is what I present to the public today.

Certainly I studied electromagnetic energy in college, yet I was never satisfied with their descriptions. To start, most descriptions are abstract, or worse, very highly mathematical. I wanted a more physical description, something that made sense as a physical object. If you could hold a ball of light in your hands, to look at it and study it, this is what you would see.

Thus, the drawings and descriptions in this book represent physical realities, physical objects, which we can comprehend in a very practical way.

Yet my concern went beyond that. Something was amiss. At the very least was the central debate of "is it a particle or is it a wave?" (I now have a definitive answer to that question, with my own insights on the process).

There were many other questions as well. For example, consider the concepts of penetration and size of the energy. Wavelength cannot correlate with size of the energy, because wavelength is a width, and the penetration size is related to diameter. You might say Amplitude then defines size, but this doesn't work either – the amplitude is irrelevant regarding penetration (as the frequency alone determines penetration size). I have now figured out these answers - as you will see in this series of chapters.

This was just one of the puzzles which concerned me over the years. There were also many other anomalies I wanted to resolve. It is only in the last year that I have figured out all of the answers.

Added note: scientists such as Bohr felt the same way about certain topics. This is what led such eminent scientists to make advancements in our understanding of science. In the same spirit, my desire for understanding anomalies in the current understanding of electromagnetic energy, led me to make further advancements in the understanding of science related to EM energy.

Yes, I understood what the texts were trying to tell me. Yes, I understood the equations (as complex as they were), and could use those equations for my own calculations. Yet I wasn't satisfied with the explanations. Something was missing.

Since college I have read about electromagnetic energy in various magazines and books. I have studied it, and I have worked with it over the years.

At the same time I became very much interested in electrical power. This became my passion. I devoted myself to studying every technology related to electrical power, I talked with experts, and I visited facilities. And in this area too I made some unique discoveries. (These books and discoveries are now published in the series "Energy Technology Explained Simply").

During this detailed study, I learned details about the generation and transmission of electrical power. This includes alternating current and voltage on the atomic scale, magnetic fields, and so much more. I also started reading the works of Tesla. This includes both his inventions and his general descriptions of phenomena.

Then suddenly in February-March of 2012 everything made sense. Within a few weeks, everything became clear. I devised new theories on the fundamental nature of electromagnetic energy. From these models and mechanisms I was able to explain all other aspects of electromagnetic energy. The new models could explain everything!

A second burst of insight came in October-December 2012. These were additional insights which incorporated additional models and additional explanations, with the previous new models, in order to explain more aspects of observed EM phenomena.

Finally, it all made sense. All characteristics of electromagnetic energy are understood. All physical properties are understood. With just a few breakthroughs in understanding, all properties of electromagnetic energy can be explained.

In your hands you have the new models. You have the guides which shows how these new models can explain the previously observed phenomenon.

In your hands you also have an easy-to-understand set of guides on electromagnetic energy. I provide you with unique analogies which will help you understand all traits of electromagnetic energy. I provide numerous drawings, so that all physical characteristics and all processes are easily understood.

You now have a complete description, an accurate understanding, of all characteristics of electromagnetic energy. I hope that this deeper understanding of electromagnetic energy will enrich your life.

M.F.

# Table of Contents
## for
### *Mysteries of Electromagnetic Energy, Part 1: Fundamental Properties of Electromagnetic Energy*

1. <u>Understanding Electromagnetic Energy</u>  <u>15</u>
   a. Introduction  15
   b. Not really a Wave  15
   c. Characteristics and Motions of Electromagnetic Energy  16

2. <u>Pulsating Energy Fields</u>  <u>19</u>
   a. Introduction  19
   b. Electromagnetic Fields  20
   c. Pulsating: Inflating and Deflating  21
   d. A Closer Look at the Pulsating Energy Fields  23
   e. Example and Description of Pulsating Electromagnetic Fields  26
   f. Frequency  29
   g. Distinct Pulsations of Electric and Magnetic Fields  30

3. <u>Creation of Wave Patterns and Particle-Wave Duality</u>  <u>33</u>
   a. Introduction  33
   b. Forward Motion  33
   c. Creation of Wave Patterns  34
   d. Frequency and Wavelength  36
   e. Amplitude  39
   f. Total Pulsating Action  40
   g. Particle-Wave Duality  41
   h. Table of Frequencies, Wavelengths, and Energies  42
   i. Energy and Frequency  45
   j. Mathematical Relationships: Energy, Frequency, and Wavelength  47

## 4. Energy and Power of EM Energy — 49
a. Introduction — 49
b. Inherent Energy of EM Burst — 49
c. Kinetic Energy of EM Burst — 50
d. Increasing Energy — 52
e. Power of EM Energy: Group Power — 54
f. Increasing the Power — 55
g. Focus and Intensity — 56

## 5. Energy Spread — 59
a. Introduction — 59
b. Group Spread of Energy Emission — 59
c. Frequency, Wavelength, Amplitude with Group Spread — 60
d. Stars and Flashlight as Group Spread — 61
e. Spread of Individual Energy Burst — 62
f. Spread of Individual Energy Burst versus Amplitude — 63

## 6. Energy Strings and General Structure of EM Burst — 67
a. Introduction — 67
b. Energy Segments — 67
c. Energy Strings in a Burst of Electromagnetic Energy — 69
d. Compare and Contrast to General String Theory — 71
e. Thickness and Length of Energy Strings — 73
f. Mass of Energy Strings — 74
g. Separating, Traveling, and Joining of Energy Strings — 76
h. Snapshot of the EM Burst — 77
i. EM Core — 78
j. Energy Strings in Emission and Absorption — 78

## Appendix Items
- Summary of Concepts, including New Theories and Models — 81

# Chapter 1
# Understanding Electromagnetic Energy

## Introduction

What exactly is a radio wave? At its most basic, a radio wave is a pulsating ball of energy that travels through space.

> A Radio Wave is a Pulsating Ball
> of Energy which Travels through Space.

Each song, speech, or other message which is broadcast through electromagnetic radiation is a distinct entity. It is separate from all other signals emitted before and after that message. Each one has a definite size, travels in its own way, and acts independently of any other signal.

> Each Signal of Electromagnetic Energy is a
> Distinct Burst of Pulsating Energy,
> with its own properties,
> and traveling in its own way through space.

## Not really a Wave

Although commonly referred to as a wave, the "radio wave" is in fact not a wave at all. The object we are discussing is a pulsating ball of energy, which can be tracked as a wave. We will discuss the specifics and explain clearly what is going on throughout this book.

For this reason it is better to use the term "electromagnetic energy" rather than radio wave. However, because of its common usage, we will use both terms throughout the book.

# Characteristics and Motions of Electromagnetic Energy

Each burst of electromagnetic energy has the following important characteristics and motions:

## 1. Distinct Object like a Baseball

Each burst of electromagnetic energy is like a baseball. It is a distinct object of specific size which can be thrown into the air. This object will fly through the air until it hits a wall or is caught by a catcher.

> A burst of Electromagnetic Energy is like a Baseball.
> It can be emitted or thrown into the air.
> It will fly through the air in a straight line,
> until it is caught, absorbed, or bounced back.

## 2. Pulsating like a heart

Each burst of electromagnetic energy exists very much like a human heart. It is a distinct object which pulsates.

Each pulse looks something like a balloon blowing up then deflating. The pulsation occurs in four directions: two directions for the "electro" part, and two directions for the "magneto" part.

Just as different animals have different heart rates so each burst of electromagnetic energy pulsates at different rates. The rate of the pulse is what determines the frequency.

The frequency of the pulse is a function of the energy level. When a burst of energy pulsates faster (greater frequency), then that burst has higher energy.

> A burst of Electromagnetic Energy pulsates like a Heartbeat.
> Each pulse looks like a balloon inflating and deflating,
> where the "balloon" is "filled" with electromagnetic energy.
>
> The rate of the pulsation is commonly known as the frequency.

## 3. Tracked as a Wave

Electromagnetic energy is not a wave, but it can be tracked as a wave. This distinction is important.

Just as Mars is not an ellipse - Mars is actually a planet which can be tracked as an ellipse - the burst of electromagnetic energy is not a wave, rather it is a pulsating ball of energy which can be tracked as a wave. Again, this distinction is important.

It is the combination of two motions - pulsating motion and traveling straight through the air - which creates the overall motion which can be tracked as a wave. All wave like properties come from the combination of those two motions.

> Electromagnetic Energy is not a wave.
> The motions of the electromagnetic energy
> can be tracked as a wave,
> but the object itself is a ball of energy, not a wave.

## 4. Spreads out like Pizza Dough

Electromagnetic energy is rarely emitted as one, singular burst. Rather, many EM bursts are usually emitted at the same time. This becomes a group of photons, sometimes referred to as "packets".

This group starts out as one entity, almost like one big ball. Yet it spreads out like pizza dough. Each group starts as a close collection of particles, then spreads out in a perfect circle.

As the group of EM bursts travel through the air, they spreads out thinner and thinner – like a very thin pizza crust.

When you spread pizza dough out too thin, some holes will appear. In the same way, the original large grouping of EM bursts will eventually spread out so thin that holes of empty space appear between each burst. This is the phenomenon which produces the night sky – enormous stars visible to us only as tiny dots – versus our Sun which is very bright because it is close.

> Electromagnetic Radiation is like Pizza Dough.
> Emission begins as a big ball of grouped EM bursts,
> then spreads out in a perfect circle
> getting thinner with each pulse
> as the individual photons spread further apart.

## Combining the Characteristics

Now let us take a look at the overall characteristics of this creature we call electromagnetic radiation.

- Each signal (such as a note of music which is broadcast over the radio) is a distinct entity unto itself. This entity is an independent object, which contains a specific amount of energy.

- Each burst of electromagnetic energy is primarily a ball of energy, similar to a baseball. Each burst can be emitted and received (just as a ball can be thrown and caught). An EM burst travels through the air in a straight line from point A to point B, until it is caught, absorbed, or reflected.

- Each burst of electromagnetic energy pulsates like a heart, with pulses that resemble balloons inflating and deflating.

- Individual bursts spreads out like pizza dough. Most emission starts as thousands of photons grouped together. As these EM bursts travel, they spread out thinner, and evenly, in a perfect circle.

Therefore, the individual EM burst is: a ball of energy, traveling through the air like a baseball, and pulsating at the same time.

Any individual burst begins as part of a grouping of identical bursts, yet as they travel through space, the individual EM bursts spread outward evenly, in the exact pattern of pizza dough spreading out.

These are the overall motions of the bursts of electromagnetic energy. All other qualities, practical effects, and applications of electromagnetic radiation come from these main characteristics, or a combination thereof.

## Future Discussions

These are the main characteristics and motions of electromagnetic energy.

I hope you like my analogies! Once I understood the true nature of electromagnetic energy, all four characteristics made sense. Many other aspects started to make sense as well. We will discuss those primary characteristics and many other concepts throughout the book.

# Chapter 2
# Pulsating Energy Fields

## Introduction

An important part of the new understanding of Electromagnetic Energy is the concept of pulsating electromagnetic fields. Therefore, in this chapter you will learn two fundamental concepts: 1) the nature of energy fields, and 2) the pulsating action of the electromagnetic energy.

A burst of electromagnetic energy is made of two fields: an electric field, and a magnetic field. These fields are lines of energies, and they flow in a particular direction. The fields always flow perpendicular to each other.

Therefore a burst of electromagnetic energy is made of criss-crossing lines of energy. A burst of electromagnetic energy looks very much like a hand-woven bag.

---

A burst of electromagnetic energy is made of criss-crossing lines of energy.

A burst of electromagnetic energy looks very much like a hand-woven bag.

---

This energy pulsates. It expands, contracts, then expands again. This physical action of pulsating electromagnetic fields explains most of the "wave" properties, including frequency, wavelength, amplitude, and much more.

When we fully understand the nature of these pulsating energy fields we can understand most other aspects of electromagnetic energy. Therefore we will examine and discuss the properties of this pulsating energy in great detail.

# Electromagnetic Fields

Introduction

Electromagnetic energy has two components: electric and magnetic. Each component exists in a "field".

What is a field? A field is energy. This energy field consists of lines of energy, which flow in a particular direction. These lines of energy can expand, contract, bend, change direction, and move in a variety of different ways.

Therefore the electric field is movement of electrical energy along a series of electrical energy lines. The magnetic field is the movement of magnetic energy along magnetic energy lines. In nature, these fields can exist separately or exist together. When they exist together, as in our burst of electromagnetic energy, the fields (energy lines) always flow perpendicular to each other.

The net result is an object that resembles a ball of yarn or hand-woven bag.

Figure 2.1: Example of EM Burst, showing EM Energy Fields

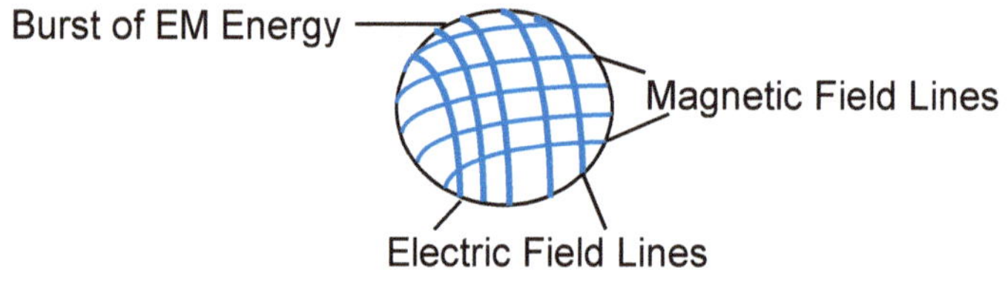

- Burst of EM Energy
- Magnetic Field Lines
- Electric Field Lines

---

A burst of electromagnetic energy is made of
an electric field and a magnetic field.

These fields are lines of energies, and flow in a particular direction.

In a burst of electromagnetic energy,
the fields flow perpendicular to each other

# Pulsating: Inflating and Deflating

Each burst of electro-magnetic radiation pulsates like a heart. This pulsating action is key to understanding many properties of EM radiation.

## The Balloon

Imagine a balloon. This balloon inflates, deflates, then inflates in the opposite direction. That is what the pulsating electric field looks like. The electric field is like a balloon, composed of electric energy. This field expands, contracts, and expands in the opposite direction – exactly like a balloon.

Similarly, there is a second balloon, perpendicular to the first. This "balloon" is composed of magnetic energy. This balloon of magnetic energy also expands, contracts, and expands in the opposite direction.

## The Hand-Woven Bag

Let us take this a step further, to a more accurate analogy. Imagine a hand-woven bag. Crumple the bag into a tight ball. Then expand the bag, stretching it out fully. Crumple the bag again, then stretch the hand woven bag in the opposite direction. There you have a pulsating action of the hand-woven bag.

The pulsating electric field is exactly the same. The electric field is composed of interwoven lines of electrical energy, much like the hand-woven bag. The electric field starts in a little ball of energy, much like the crumpled hand-woven bag. Then the electric field expands out, reaches its fullest extent, then retreats, and expands out in the opposite direction (exactly like the hand-woven bag where we stretch, crumple, and stretch out again).

Of course the magnetic field exists as a second "hand-woven bag", composed of interwoven lines of magnetic energy. This bag starts as crumpled up, then stretches out, crumpled again, then stretches out in the opposite direction.

---

The Pulsing of the Electric and Magnetic Fields is like
two perpendicular balloons inflating and deflating,
or two hand-woven bags stretching and compacting.

---

## Pulsating Electromagnetic Fields

Beyond the analogies, the specific actions of the pulsating electromagnetic fields is as follows: Each field pushes outward, then shrinks back to a center point, and then pushes outward in the opposite direction. This series of steps creates one full pulse. Then the cycle repeats.

Note that you have two energy fields individually pulsating at the same time. However, each field pulsates outward at the same frequency, and each field pulsates to the same maximum distance.

Note also that the fields always exist perpendicular to each other. For example the electric field may push upward while the magnetic field pushes to the left. Then the electric field would push downward while the magnetic field pushes to the right.

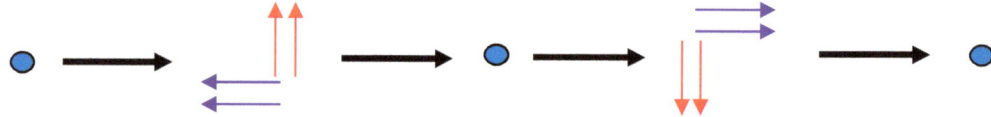

Most important, note that the two fields combine. This merging of fields creates our special balloon or bag shape in each direction. The details of this physical entity will be described in detail below.

# A Closer Look at the Pulsating Energy Fields

Introduction

Here we will take a closer look at the pulsating energy fields. You will begin to understand what the pulsating electromagnetic fields really look like. Most significantly, you will understand the pulsing energy fields on a physical level, rather than abstractly as most texts try to describe it.

Note also that this is one of the major insights I have developed regarding electromagnetic energy.

We will go through the process step by step, from the abstract to the physical, so that you can truly understand the physical qualities of electromagnetic energy.

Vectors Showing Paths

Remember that there are two energy fields, electric and magnetic. Also remember that these fields run perpendicular to each other. Therefore it is common in the physics world to show these fields as perpendicular arrows:

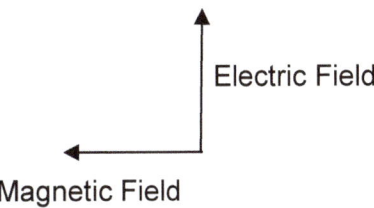

What this simple diagram really tells you is: 1) all the electric fields in the energy burst are flowing upward, and 2) all the magnetic fields in the energy burst are flowing to the left.

Energy Between the Vectors

However, what most people don't realize is that there is actually a volume on energy between the vectors:

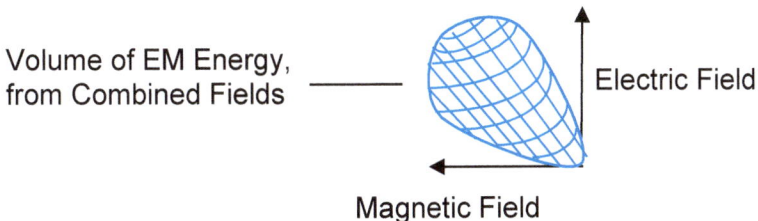

We will go through this step by step so that you can see how this volume of electromagnetic energy is created.

## Energy Around The Vectors

The single vector is somewhat misleading. The energy fields are broader than these vector lines indicate. Energy fields are not these narrow straight lines. Instead, the fields exist around these lines.

It is true that the vector shows direction. It is also true that there are invisible lines at the center of each field, and those lines are the strongest areas. Yet it is also true that energy flows around the vectors as well.

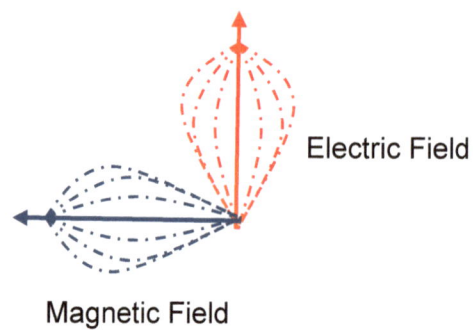

## Energy Fields Merging

These energy fields actually begin to merge between the vectors

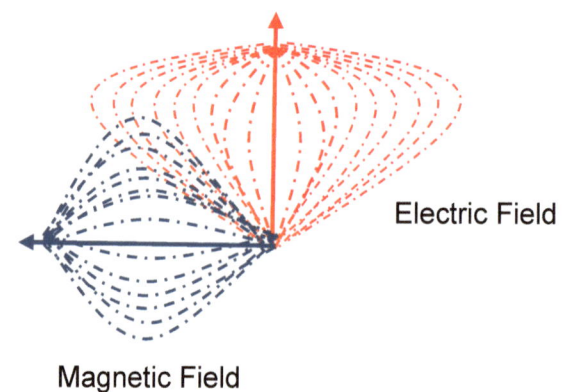

## Merging Fields creates a Unique Identity

The merging electric field and magnetic field creates a unique identity, a separate entity of "electro-magnetic energy".

## The Volume of Energy Between The Vectors

There is actually a volume of energy between those two vectors. This is the physical entity of the electromagnetic energy pulse.

As stated above:

a. Each vector shows the path of the energy field during any pulsation.

b. The energy field is strongest along the invisible field lines.

c. However the field is much broader than a simple line. The energy field is surrounds the central line.

d. The energy fields of the electric field and the magnetic field are broad enough to merge.

e. Their merging creates a new energy: the combined "electro-magnetic energy" with its own "electro-magnetic field"

The result is therefore a volume of space where the two fields interact. This is the merging of electric field and magnetic field. Within that volume of space is the majority of electro-magnetic energy. This volume of space generally is shaped like a balloon or a hand-woven bag.

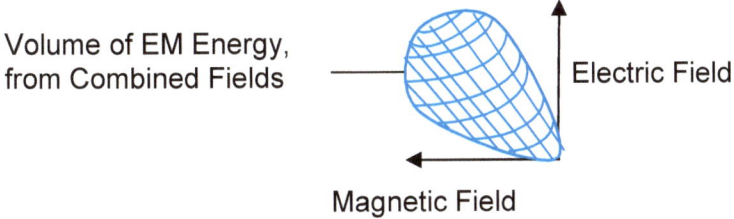

Volume of EM Energy, from Combined Fields — Electric Field

Magnetic Field

> The physical entity of the electromagnetic pulse is a volume of space which contains the merged electric and magnetic fields.

# Example and Description of Pulsating Electromagnetic Fields

Now we can truly understand the pulsation of the electromagnetic energy burst. Below is an example and a description of the process:

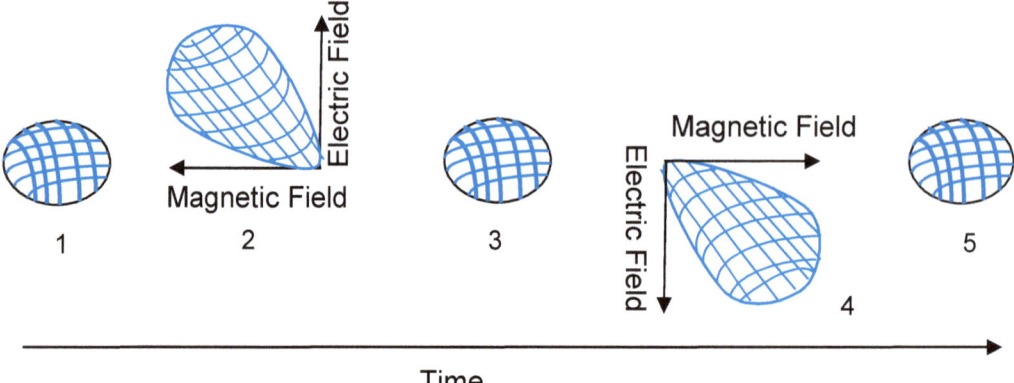

The Pulsation Motion of Electromagnetic Energy consists of the following steps:

1. Energies Begin as Compact Sphere
2. Combined Energy Fields Extend One Direction
3. Energies Retreat to Compact Sphere
4. Combined Energy Fields Extend Opposite Direction
5. Energies Retreat to Compact Sphere

This process repeats essentially forever.

## Brief Description of Pulsation Process

In brief the process is as follows: The electromagnetic energy burst begins as compact sphere. As the burst travels forward it begins to expand in one direction. When the burst has reached its maximum expansion, it begins to contract, returning to its original sphere state.

Then the burst begins to expand in the opposite direction. The burst expands to its maximum in that opposite direction. After reaching the maximum, the energy begins to contract. Again the sphere reaches its original sphere state.

That process is one complete pulsation of the electromagnetic energy burst. This pulsation cycle continues repeatedly forever (or until absorbed by a material).

Detailed Description of Pulsation Process

Combing the basic pulsation steps above with our understanding of electric fields and magnetic fields, we can be more accurate with our description of the pulsation process:

1. The energy burst begins as a compact sphere of electromagnetic energy. Inside this sphere are two energy fields: electric field and magnetic field.

2. As soon as the burst is emitted, it begins to expand.
   a. Specifically this means that the electric field flows one direction, and the magnetic field flows in a perpendicular direction.
   b. These energy fields merge and combine in the diagonal between the two main field directions. This combination produces a new entity: the electromagnetic field.
   c. A volume of space is created where the two energy fields merge. This volume of space is shaped like a balloon, and has the appearance of a hand-woven bag.
   d. This volume of space contains and is composed of the merged electric and magnetic fields. This is the physical entity of the electromagnetic energy during the expansion phase of the pulsation cycle.
   e. The main fields push out in each direction, which causes the balloon shaped volume of EM energy to expand in a diagonal direction.
   f. At some point the electric field and magnetic field have pushed energy out to a maximum position. The diagonal balloon of EM energy is also pushed to its maximum position.
   g. All energy fields begin to retreat. The electric field stops pushing as much, the magnetic field stops pushing as much, and the balloon-bag of EM energy begins to collapse.

3. The burst of electromagnetic energy returns to its original sphere state.

4. The energy burst begins to expand, in the opposite directions. All processes are the same, but occur in opposite directions horizontally, vertically, and diagonally.

    a. The electric field flows 180 degrees opposite to the direction it did before. The magnetic field also flows 180 degrees opposite to the direction it did before. The electric and magnetic fields are still perpendicular to each other.

    b. As each energy field grows, these energy fields merge and combine in the diagonal between the two main field directions. This combination produces the entity of the electromagnetic field.

    c. A volume of space is created where the two energy fields merge. As before, this volume of space is shaped like a balloon, and has the appearance of a hand-woven bag. However, it is pointed in the exact opposite diagonal from before.

    d. As above, this volume of space contains the merged electric and magnetic fields. This is the physical entity of the electromagnetic energy during the second expansion phase of the pulsation cycle.

    e. The main fields push out in each direction, which causes the balloon shaped volume of EM energy to expand in a diagonal direction.

    f. At some point the electric field and magnetic field have pushed energy out to a maximum position. The diagonal balloon of EM energy is also pushed to its maximum position.

    g. All energy fields begin to retreat. The electric field stops pushing as much, the magnetic field stops pushing as much, and the balloon-bag of EM energy begins to collapse.

5. The burst of electromagnetic energy returns to its original sphere state.

This completes one full pulsation cycle of an individual burst of electromagnetic energy.

Note that I have spent such a significant amount of time on this topic because it is a central concept to everything about electromagnetic energy. Understanding the nature of electromagnetic fields and the exact mechanism of pulsation will help you understand many things about electromagnetic energy which we will discuss in the future.

## Frequency

### Introduction

Now that we understand the pulsation cycle of the energy fields, we can easily understand the concept of frequency: The frequency of a burst of electromagnetic energy is simply the rate at which the pulse goes through one complete cycle.

> The Frequency of an Electromagnetic Energy Burst is the time it takes for the energy field to complete one full cycle of pulsing in all directions.

### Frequency and the Cycle

A cycle begins with the energy field is compacted. The energy expands, ultimately to its largest position, then shrinks again to the compacted state, at which point it expands in the opposite direction, and shrinks back once again to its compacted state. That is one full cycle of pulsing. Therefore, the time it takes for the energy burst to complete one full cycle the frequency of our energy burst.

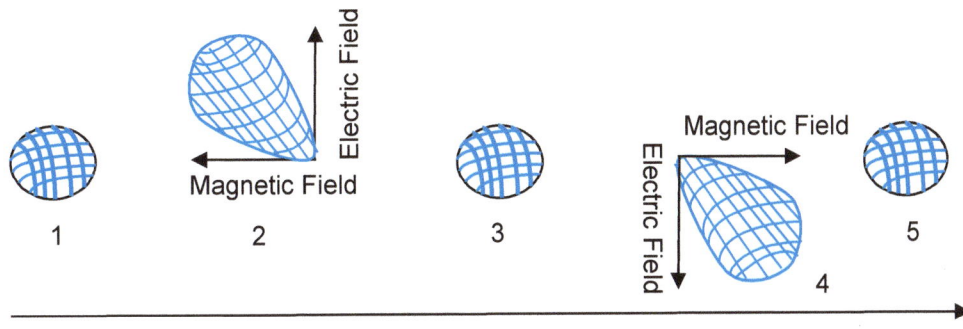

Cycles per Second

Frequency of electromagnetic energy is measured in number of cycles per second. This is the number of times our burst of electromagnetic energy completes full pulsation sequence in each second.

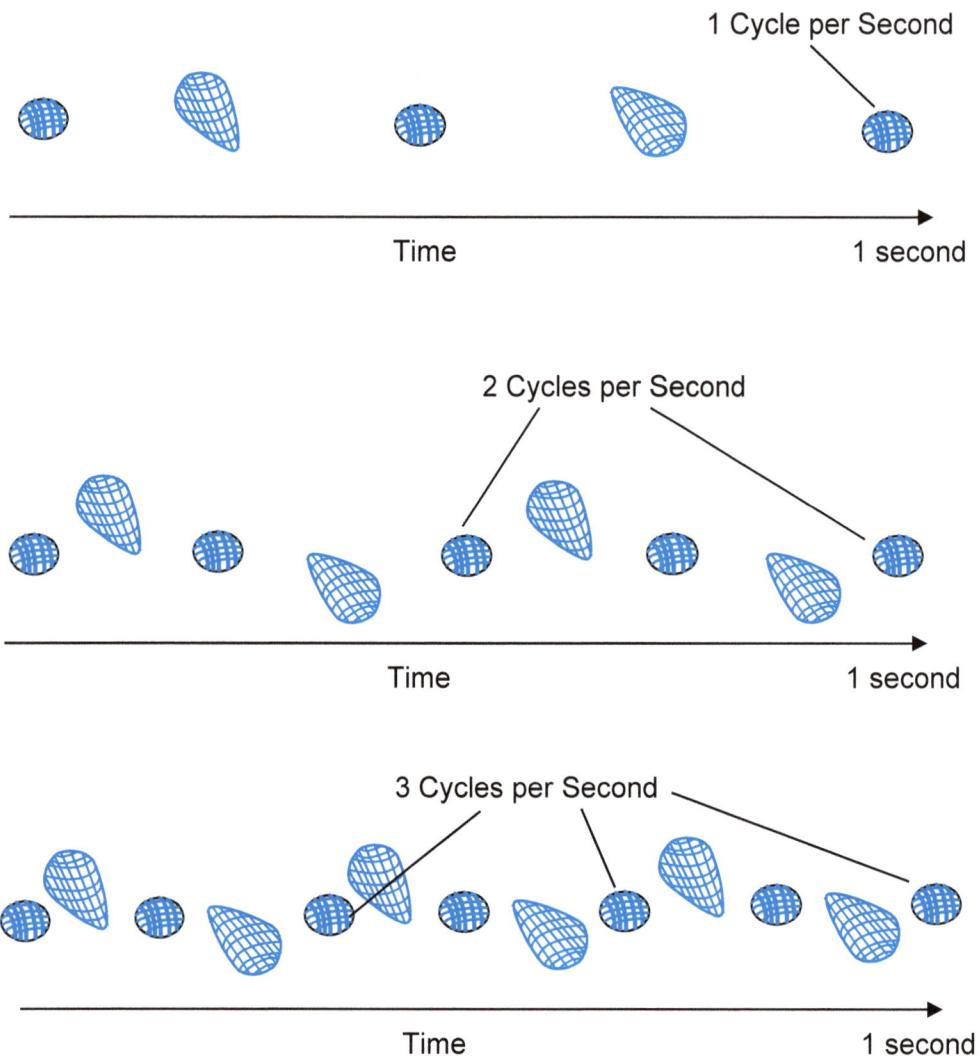

If a burst of electromagnetic energy pulsates very quickly, then it will complete several cycles – perhaps millions of cycles - in each second. If a burst of electromagnetic energy pulsates slowly, then it will take a long time for the burst to complete its cycle – perhaps only a few cycles, or only one cycle – in a second.

# Distinct Pulsations of Electric and Magnetic Fields

## Introduction
At this point I would like to emphasize that we actually have two distinct energy fields. The fields are related, yet are in some ways separate.

## Similarities of Both Fields Related to Pulsation
Both the electric field and the magnetic field will pulsate at the exact same frequency. This is because the strings are in the same general region of space, and the cause for the rate pulsation for one field will be the cause for the rate of pulsation of the other field. (This will be described in detail in later chapters).

## Differences in Fields as Related to Pulsation
On the other hand, each field has its own energy strings. Therefore, the strength of each field depends on that field alone. More specifically, the energy of the field and the amplitude of the pulsation of that field depends on that field alone.

For example, the strength of the electric field can be much greater than the strength of the magnetic field. As another example, the amplitude of the magnetic field (how far the strings flow out before coming back) can me much further than the amplitude of the electric field.

Again, the frequency of pulsation is exactly the same for both fields (because they share the same space). Yet the strength and amplitude of each field depend only on the dimensions of the strings for that particular field.

## Individual Fields versus Merged Fields
In reality, our pulsation process involves four dimensions. For example, our EM burst will pulsate upward (for electric field) and left (for magnetic field). Then both fields retreat. Then our EM burst will pulsate the opposite directions, such as downward (for electric field) and right (for magnetic field).

Thus, in reality our EM burst is pulsating two perpendicular balloon shapes out, then retreat, then two perpendicular balloon shapes outward in the opposite direction.

For simplicity in our discussions it is best of focus on just one pulsation, rather than both. This is why I have chosen to focus on the region where the two fields overlap, where they create the "merged" electromagnetic field.

Thus, throughout this book I will focus on "one" pulsation – the pulsation of the "merged" electric field and magnetic field. This will simplify all of our illustrations and all of our discussions.

This also works because both fields have the exact same frequency. Therefore all of our discussions on cause of frequency can be done using the region of space where the fields are merged.

However, there are always two fields, pulsating perpendicular to each other, and there are times where I will focus on just one field, such as the electric field alone.

Yet there is a region where the two fields are merged (as I use in my diagrams throughout this book), and most of our discussions will be focused on the single region of the merged fields.

# Chapter 3
# Creation of Wave Patterns and Particle-Wave Duality
# (Also Discussion of Frequency, Waves, Amplitude)

## Introduction

As stated many times – because this is an important concept – a burst of electromagnetic energy is not a wave. Rather, a burst of electromagnetic energy is a sphere of energy. Its motion can be tracked as a wave, but physically it is a sphere not a wave.

Now that we understand the process of pulsating electromagnetic fields, we are able to understand the wave pattern: The creation of the wave pattern is a combination of the forward motion of the energy burst, and the repeated pulsing of the electromagnetic fields.

The following sections will explain the wave pattern, and all related concepts, in greater detail.

## Forward Motion

The pulsing motion is only one part of the situation which creates the wave characteristics. The other part is the forward motion. Therefore let us revisit the concepts behind the forward motion of electromagnetic radiation.

Remember our first concept of electromagnetic radiation: a discrete burst of energy much like a baseball, which can be emitted, received, and fly through the air in a straight line. It is this straight line we will focus on now.

Figure 3.1

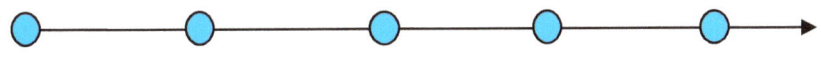

Forward Motion of Energy Burst

Any burst of electromagnetic radiation flies in a straight line. It does not swerve or travel in a wave. Yes, it can be deflected. Yes, it can be bent by certain gravitational pulls. However, I want to emphasize that the burst itself does NOT travel in a wave.

The burst does not travel in a zig-zag pattern or wander around aimlessly. A burst of EM radiation travels in an absolute straight line until it is deflected or received.

Note that the speed of forward motion is the same for all bursts. This is one of the few constants of electromagnetic radiation. All types of bursts, regardless of energy level or pulsating frequency, will travel forward at the same speed.

## Creation of Wave Patterns

### Overview

Remember what we said in our new understanding of electromagnetic energy: "radio waves" are *not* waves. Electromagnetic radiation is in fact a burst of pulsating energy which is traveling through space. That is the fundamental characteristic of electromagnetic radiation.

There are wave properties and wave characteristics, but the EM radiation itself is not a wave. It doesn't exist as a wave, nor does it travel as a wave. This is an important distinction.

So if electromagnetic radiation is not actually a wave or travel as a wave, then how does it have wave properties? We will answer this question in the next several sections.

In brief, it is the combination of the pulsing action and the forward motion of the electro-magnetic radiation which creates a wave pattern. As viewed from the side (or from above), any burst of electromagnetic radiation can appear to have a unique wave pattern as it travels through space. We will demonstrate this in the diagrams below.

> The Wave Pattern is created by the combination of the pulsating sequence and the forward motion of the energy sphere.

Figure 3.2

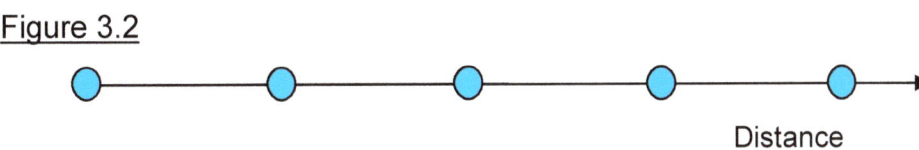

Distance

Forward Motion of Energy Burst

Figure 3.3

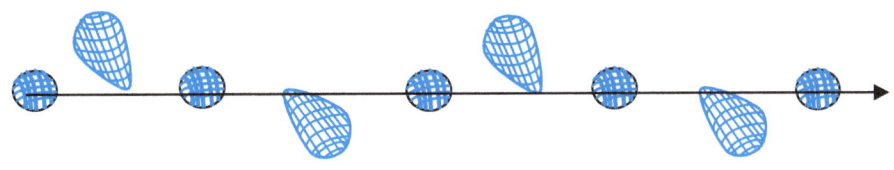

Pulsation Cycle         Time and Distance

Figure 3.4

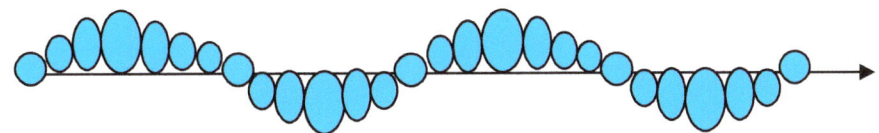

Pulsation Cycle combined with Forward Motion,
as viewed from the side.

Figure 3.5

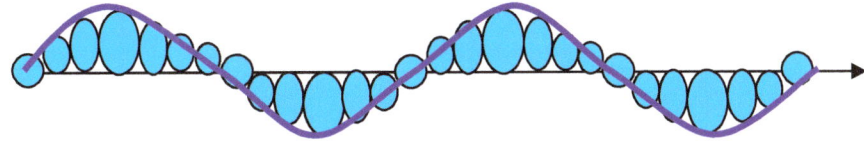

Creation of the Wave Pattern:
A combination of pulsation cycle and forward motion
(as viewed from the side) can be tracked as a Wave.

# Frequency and Wavelength

## Overview
Now that we understand the creation of the wave pattern, we can begin to understand all other aspects of the wave.

## Frequency
As we said earlier, the frequency is simply the rate of pulsation. Each burst of electro-magnetic radiation goes through a cycle of pulsing. Each burst goes through a series of pulsating steps: stretching outward, retreating, compacting, then stretching out into the opposite direction, retreating, and compacting again. These events comprise one complete cycle of pulsation. The time it takes for a burst of EM radiation to complete this cycle is its frequency.

## Speed of Forward Motion
All bursts of electro-magnetic radiation, regardless of frequency, travel forward at the exact same rate. This is an important concept when working with wavelengths.

## Wave Pattern
A wave pattern is created by the combination of the pulsating motion and the forward motion. As the burst pulsates, while at the same time travels forward in space, a wave pattern is created. This wave pattern is best seen by viewing the traveling wave from the side.

## Wavelength
The wavelength is the distance traveled by the burst of energy while completing one full cycle of pulsing.

Figure 3.6

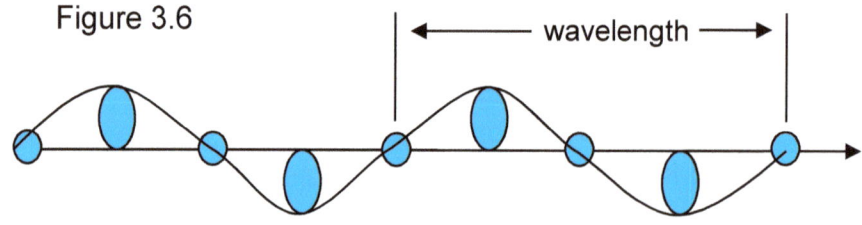

## Relationship Between Frequency and Wavelength

The exact size of the wavelength is related to the frequency of the pulse. A slower pulse (lower frequency) results in a longer wavelength. Similarly, a faster pulse (greater frequency) results in a shorter wavelength. This relationship exists because the forward motion is the exact same rate for all bursts of energy, whereas the pulse frequencies can vary quite a bit.

For example, consider a burst of energy with a very slow pulse. It seems to take forever to complete its pulsing cycle. Yet it is flying through the air at the same speed as all other bursts of energy. Therefore, by the time the burst of energy finally completes its full cycle of pulsing, this energy ball has traveled a long, long distance through the air. Therefore we have a longer wavelength.

Similarly, consider a burst with a very fast pulse rate. This burst of energy goes through its complete cycle very quickly. It is practically hyperactive. Yet it is flying through the air at the same speed as all other bursts of energy. Therefore the burst will be traveling a very short distance, perhaps barely noticeable, by the time the energy ball completes its cycle. Therefore we have a shorter wavelength.

# Figure 3.7: Comparison Examples Showing Relationship of Frequency and Wavelength

Let us use the same three examples we used previously for showing comparisons of pulsation frequency. Using these examples, we can also show comparative wavelengths. And of course, we can see the relationship of frequency to wavelength.

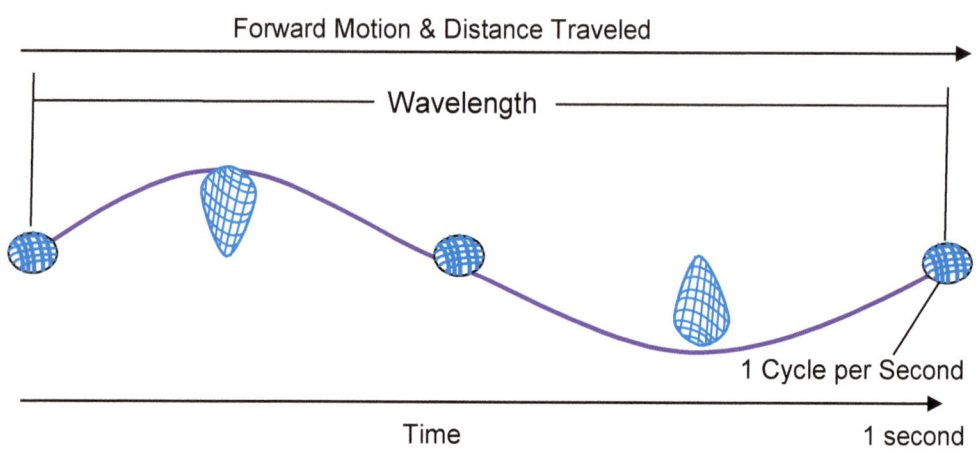

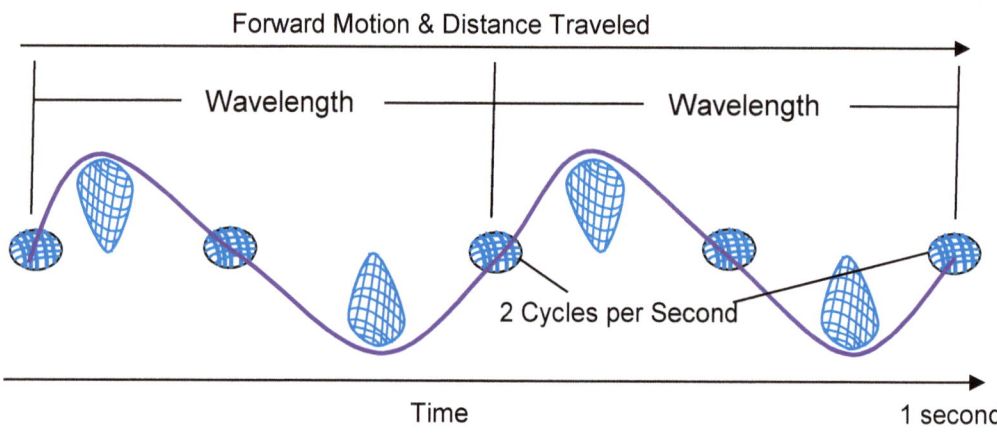

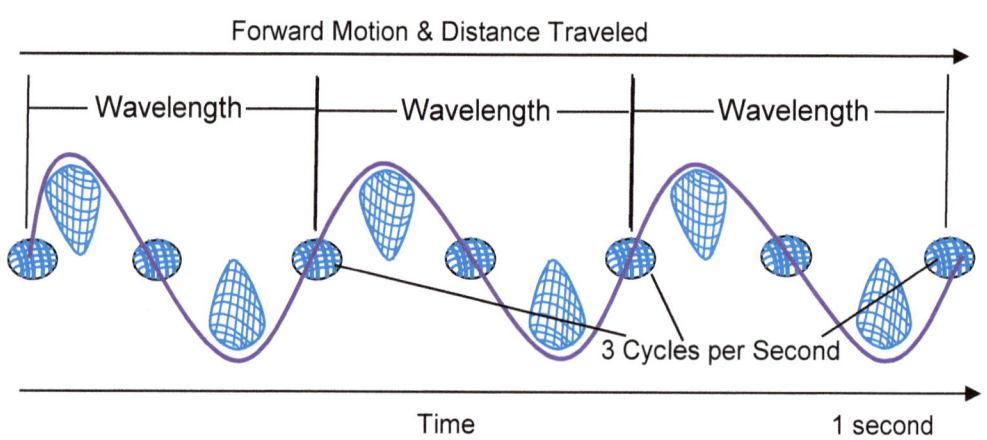

# Amplitude

## Amplitude Basics

Now that we understand how the wave pattern is created, we can understand the Amplitude of the wave. The amplitude of a wave is simply the distance outward which the electromagnetic field stretches.

Remember our analogy of the hand-woven bag? The amplitude would be how far the hand-woven bag stretches out. The fullest distance which the bag stretches out before retreating is the amplitude.

<u>Figure 3.7</u>

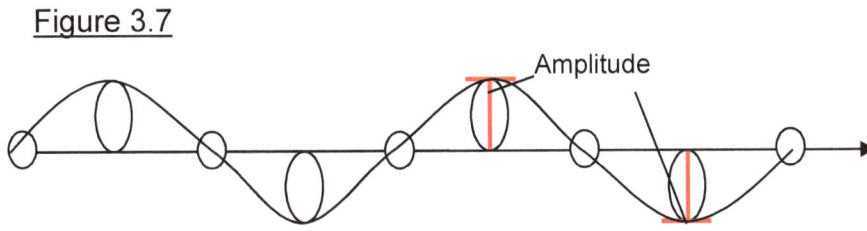

> The Amplitude is the fullest stretch of the electromagnetic fields, in either direction, during the pulsation cycle.

## Total Pulsating Action

The following is an example of everything together. In this simple illustration you can see the EM Energy Fields, the Pulse Cycle, the Amplitude, the Wave Pattern, and the Wavelength.

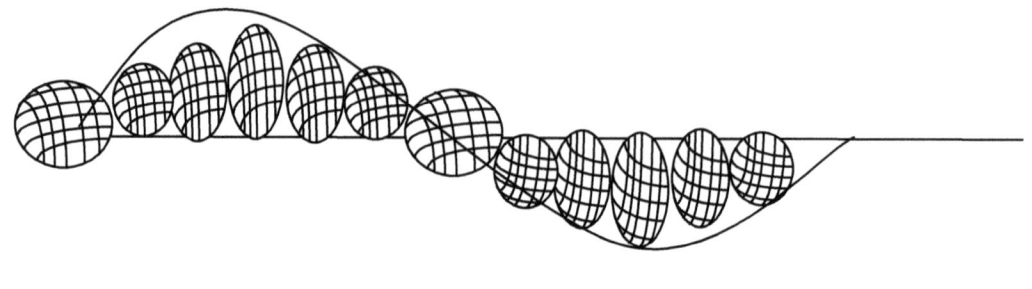

# Particle-Wave Duality

Now we can understand Particle-Wave Duality for electromagnetic energy:

1. Particle: A burst of electromagnetic energy is foremost a particle. It is a sphere of electric and magnetic energy fields. This sphere of energy fields flies forward through space similar to a baseball.

2. Wave: The burst of electromagnetic energy is not a wave. Rather, the motion of the burst can be tracked as a wave.
The wave motion of the EM burst is created as a combination of two motions: the pulsation motion and the forward motion.

2a. The EM burst pulsates as it travels. Each energy field pulsates outward in one direction, then collapses inward, and then pulsates outward in the opposite direction. This pulsation repeats continuously.
2b. At the same time, the EM burst is flying through the air. This looks very much like a baseball as it flies across a field.
2c. The pulsation motion and the forward motion exist together, at all times. When you look at these combined motions from the side, and trace the path, you will indeed see a wave.

3. Particle-Wave Duality: Therefore, again, the Particle-Wave Duality for electromagnetic energy can be explained as follows: any burst of EM energy is primarily a particle, yet the combined motions of the particle (the pulsation and forward trajectory) can be tracked as a wave.

# Table of Frequencies and Wavelengths for Common Ranges of EM Energy

## Introduction

In the table below you will see a full list of categories of electromagnetic energies, with the corresponding frequency, wavelength, and energy value range for each category.

## General Notes about the Table

1. Because the electromagnetic energy spectrum is so broad, it is helpful to divide it into general categories. Therefore, the common names listed in the table are general category names, which are commonly used by the general public.

2. However, also remember that we are working with a continuous spectrum. There are no absolute dividing lines from one category to the other. The values of each range, whether for energy, frequency, or wavelength, are simply useful cut off points for comparisons.

This also extends to detectors (including various animals). The absolute range of the energies which the animal or device can detect can vary. (For example, some humans can see colors extending into the ultra violet which other humans cannot. Therefore, even among humans, the range of detectable frequencies is not an absolute).

3. Also note that the values have been rounded, even among an individual row. For example, a value of "$2.0 \times 10^{-22}$ Joules" in the Energy column has an actual value of "$9.93 \times 10^{-4}$ meters". However, for simplification we round the meters value to $1.0 \times 10^{-3}$.

This table is mostly as a guide, not for engineering. Therefore such rounding is useful as comparisons.

4. Typical lists of categories of electromagnetic energy start with wavelength, then frequency, then energy. However, energy is really the primary value. This determines the frequency, which also then determines the wavelength.

However, for the purposes of having a table beginning with a value of "1.0" we can begin with Frequency. Thus, the lowest possible frequency possible is 1.0 Hz. The other values of the table are established from there.

## Categories of Electromagnetic Energy, with Frequency and Energy

| Common Name | Frequency (Hz) Smallest to Largest | Wavelength (meters) Largest to Smallest | Energy (Joules) Smallest to Largest |
|---|---|---|---|
| Low Frequency | 1.0 Hz to 3.0 x$10^6$ Hz | 2.9 x$10^8$ m to 1 x$10^2$ m | 6.6 x$10^{-34}$ Joules to 2.0 x$10^{-27}$ Joules |
| Radio Waves | 3.0 x$10^6$ Hz to 3.0 x$10^9$ Hz | 1.0 x$10^2$ m to 1 x$10^{-1}$ m | 2.0 x$10^{-27}$ Joules to 2.0 x$10^{-24}$ Joules |
| Microwave | 3.0 x$10^9$ Hz to 3.0 x$10^{11}$ Hz | 1.0 x$10^{-1}$ m to 1 x$10^{-3}$ m | 2.0 x$10^{-24}$ Joules to 2.0 x$10^{-22}$ Joules |
| Infrared | 3.0 x$10^{11}$ Hz to 4.0 x$10^{14}$ Hz | 1.0 x$10^{-3}$ m to 7 x$10^{-7}$ m | 2.0 x$10^{-22}$ Joules to 3.0 x$10^{-19}$ Joules |
| Visible Light | 4.0 x$10^{14}$ Hz to 7.5 x$10^{14}$ Hz | 7.0 x$10^{-7}$ m to 4 x$10^{-7}$ m | 3.0 x$10^{-19}$ Joules to 5.0 x$10^{-19}$ Joules |
| Ultra Violet | 7.5 x$10^{14}$ Hz to 3.0 x$10^{16}$ Hz | 4.0 x$10^{-7}$ m to 1 x$10^{-8}$ m | 5.0 x$10^{-19}$ Joules to 2.0 x$10^{-17}$ Joules |
| X-Ray | 3.0 x$10^{16}$ Hz to 3.0 x$10^{19}$ Hz | 1.0 x$10^{-8}$ m to 1 x$10^{-11}$ m | 2.0 x$10^{-17}$ Joules to 2.0 x$10^{-14}$ Joules |
| Gamma Ray | 3.0 x$10^{19}$ Hz to 3.0 x$10^{21}$ Hz | 1.0 x$10^{-11}$ m to 1.0 x$10^{-13}$ m | 2.0 x$10^{-14}$ Joules to 2.0 x$10^{-12}$ Joules |
| Cosmic Ray | faster than 3.0 x$10^{21}$ Hz | less than 1.0 x $10^{-13}$ m | greater than 2.0 x$10^{-12}$ Joules |

| Common Name | Frequency (Hz) (Smallest to Largest) | Wavelength (meters) (Largest to Smallest) | Energy (Joules) (Smallest to Largest) |
|---|---|---|---|
| Low Frequency Waves | $1.0$ Hz to $3.0 \times 10^6$ Hz | $2.9 \times 10^8$ to $1 \times 10^2$ meters | $6.6 \times 10^{-34}$ to $2.0 \times 10^{-27}$ Joules |
| Radio Waves | $3.0 \times 10^6$ Hz to $3.0 \times 10^9$ Hz | $1.0 \times 10^2$ to $1 \times 10^{-1}$ meters | $2.0 \times 10^{-27}$ to $2.0 \times 10^{-24}$ Joules |
| Microwave | $3.0 \times 10^9$ Hz to $3.0 \times 10^{11}$ Hz | $1.0 \times 10^{-1}$ to $3.0 \times 10^{-3}$ meters | $2.0 \times 10^{-24}$ to $2.0 \times 10^{-22}$ Joules |
| Infrared | $3.0 \times 10^{11}$ Hz to $4.0 \times 10^{14}$ Hz | $1.0 \times 10^{-3}$ to $7 \times 10^{-7}$ meters | $2.0 \times 10^{-22}$ to $3.0 \times 10^{-19}$ Joules |
| Visible Light | $4.0 \times 10^{14}$ Hz to $7.5 \times 10^{14}$ Hz | $7.0 \times 10^{-7}$ to $4 \times 10^{-7}$ meters | $3.0 \times 10^{-19}$ to $5.0 \times 10^{-19}$ Joules |
| Ultra Violet | $7.5 \times 10^{14}$ Hz to $3.0 \times 10^{16}$ Hz | $4.0 \times 10^{-7}$ to $1 \times 10^{-8}$ meters | $5.0 \times 10^{-19}$ to $2.0 \times 10^{-17}$ Joules |
| X-Ray | $3.0 \times 10^{16}$ Hz to $3.0 \times 10^{19}$ Hz | $1.0 \times 10^{-8}$ to $1 \times 10^{-11}$ meters | $2.0 \times 10^{-17}$ to $2.0 \times 10^{-14}$ Joules |
| Gamma Ray | $3.0 \times 10^{19}$ Hz to $3.0 \times 10^{21}$ Hz | $1.0 \times 10^{-11}$ to $1.0 \times 10^{-13}$ meters | $2.0 \times 10^{-14}$ to $2.0 \times 10^{-12}$ Joules |
| Cosmic Ray | faster than $3.0 \times 10^{21}$ Hz | smaller than $1.0 \times 10^{-13}$ meters | greater than $2.0 \times 10^{-12}$ Joules |

# Energy and Frequency

## Introduction

There is also a relationship between frequency of the pulse and overall energy of the electromagnetic radiation. Specifically, a burst of electromagnetic energy which pulsates faster will have more energy. Conversely, a burst of electromagnetic energy which pulsates slower will have lower energy.

## Cause and Effect is Reversed

I have reversed the cause and effect for frequency and energy, which really makes much more sense. Most people say that a wave a certain frequency, and therefore it has a certain energy. I reverse this. I say that the electromagnetic energy burst has an inherent amount of energy to begin with. Then, it is this inherent amount of energy which will cause a specific frequency of pulsation.

Also note that although this may seem like a simple idea, it has profound effects. The details will be explained and demonstrated in later chapters.

## Kinetic Energy and Frequency of Pulsation

Every burst of electromagnetic energy is created with an inherent amount of energy. The factors which make this energy will be described in the next chapter. For now we will focus on one type of energy: the kinetic energy due to pulsation.

Kinetic energy is the energy of motion. Any object in motion, whether the forward movement of a car or the vibrational motion of a molecule, has kinetic energy.

Also, the amount of motion corresponds directly with the amount of kinetic energy. You are aware of this concept whenever you are physically active. To run faster or to be more agile in sports requires a greater investment of your energy. You have more kinetic energy when you produce those faster motions. Stated another way, the words "faster" and "slower" are descriptive words for motion, yet the value of kinetic energy are numbers which will quantify that motion into exact amounts.

Furthermore, the exact amount of motion corresponds directly with the exact amount of kinetic energy. For example, if a molecule vibrates faster by factor of 10, then the amount of kinetic energy, by definition, also increases by a factor of 10.

Regarding our burst of electromagnetic energy, we know that each burst will pulsate. This pulsation is a type of motion, and therefore this burst has (by definition) some kinetic energy. Furthermore, when a burst pulsates faster, it has more kinetic energy (by definition). When a burst pulsates slower, it is lower kinetic energy (by definition).

Therefore, a burst which pulsates at a faster frequency has higher kinetic energy. Conversely, a burst which pulsates at a slower frequency has lower kinetic energy. This is the basic relationship between frequency and kinetic energy in any frequency (or wavelength) of electromagnetic energy.

Again, the details of pulsation, and how the pulsation is caused by inherent energy, will be demonstrated and explained in later chapters.

## Table of Electromagnetic Energy

As stated above, the energy comes first. Thus, the electromagnetic energy burst has an inherent amount of energy. From this inherent energy is created the frequency of pulsation. Then, due to the motion of the burst, which includes the specific frequency, a particular wavelength is observed.

Therefore, the more accurate ordering of EM categories would be Energy, then Frequency, then Wavelength, as seen in the table below.

| Common Name | Energy (Joules) (Smallest to Largest) | Frequency (Hz) (Smallest to Largest) | Wavelength (meters) (Largest to Smallest) |
|---|---|---|---|
| Low Frequency | $6.6 \times 10^{-34}$ to $2.0 \times 10^{-27}$ Joules | 1.0 to $3.0 \times 10^{6}$ Hz | $2.9 \times 10^{8}$ to $1 \times 10^{2}$ m |
| Radio Waves | $2.0 \times 10^{-27}$ to $2.0 \times 10^{-24}$ Joules | $3.0 \times 10^{6}$ to $3.0 \times 10^{9}$ Hz | $1.0 \times 10^{2}$ to $1 \times 10^{-1}$ m |
| Microwave | $2.0 \times 10^{-24}$ to $2.0 \times 10^{-22}$ Joules | $3.0 \times 10^{9}$ to $3.0 \times 10^{11}$ Hz | $1.0 \times 10^{-1}$ to $1 \times 10^{-3}$ m |
| Infrared | $2.0 \times 10^{-22}$ to $3.0 \times 10^{-19}$ Joules | $3.0 \times 10^{11}$ to $4.0 \times 10^{14}$ Hz | $1.0 \times 10^{-3}$ to $7 \times 10^{-7}$ m |
| Visible Light | $3.0 \times 10^{-19}$ to $5.0 \times 10^{-19}$ Joules | $4.0 \times 10^{14}$ to $7.5 \times 10^{14}$ Hz | $7.0 \times 10^{-7}$ to $4 \times 10^{-7}$ m |
| Ultra Violet | $5.0 \times 10^{-19}$ to $2.0 \times 10^{-17}$ Joules | $7.5 \times 10^{14}$ to $3.0 \times 10^{16}$ Hz | $4.0 \times 10^{-7}$ to $1 \times 10^{-8}$ m |
| X-Ray | $2.0 \times 10^{-17}$ to $2.0 \times 10^{-14}$ Joules | $3.0 \times 10^{16}$ to $3.0 \times 10^{19}$ Hz | $1.0 \times 10^{-8}$ to $1 \times 10^{-11}$ m |
| Gamma Ray | $2.0 \times 10^{-14}$ to $2.0 \times 10^{-12}$ Joules | $3.0 \times 10^{19}$ to $3.0 \times 10^{21}$ Hz | $1.0 \times 10^{-11}$ to $1.0 \times 10^{-13}$ m |
| Cosmic Ray | greater than $2.0 \times 10^{-12}$ Joules | faster than $3.0 \times 10^{21}$ Hz | less than $1.0 \times 10^{-13}$ m |

# Mathematical Relationships Between Energy, Frequency, and Wavelength

## Introduction

The mathematical relationships between energy, frequency, and wavelength for electromagnetic energy have been known. The simplest set of mathematical relationships are as follows.

## 1. Frequency and Energy

The following three factors are related mathematically: Energy, Frequency, and Plank's Constant.

    In words:    Energy = Frequency x Plank's Constant
    In units:     # Joules = # Hz x 6.626 x $10^{-34}$ Joules/second

or

    In words:    Frequency = Energy / Planck's Constant
    In units:     # Hz = # Joules / 6.626 x $10^{-34}$ Joules/second

## 2. Frequency and Wavelength

The following three factors are related mathematically: Frequency, Wavelength, and the Speed of Light.

    In words:    Frequency x Wavelength = Speed of Light
    In units:     # Hz x # Meters = 299,792,458 m/s (in a vacuum)

or:

    In words:    Wavelength = Speed of Light / Frequency
    In units:     # of Meters = 299,792,458 m/s / # of Hz

## 3. Energy and Wavelength

The following four factors are related mathematically: Energy, Wavelength, Plank's Constant, and Speed of Light.

In words: Energy x Wavelength = Plank's Constant x Speed of Light
In units:  # Joules x # Meters = 6.6 x $10^{-34}$ J/sec x 2.9 x$10^{8}$ m/s

or:

In words: Wavelength = [Plank's Constant x Speed of Light] / Energy
In units: # Meters = [6.6 x $10^{-34}$ J/sec x 2.9 x$10^{8}$ m/s] / # Joules

# Chapter 4
# Energy and Power of EM Energy

## Introduction

Every emission of electromagnetic energy has a certain amount of energy, and a certain amount of power. In this chapter we will discuss the exact nature of energy and the exact nature of power as exists for electromagnetic energy.

In brief, the energy of the energy of the electromagnetic energy burst depends on the energy strings. We will find that the specific number of strings, the thickness of the strings, and the length of those strings will create different amounts of energy, and different practical results.

We will also introduce the concept of group emission, where multiple identical photons are emitted at the same time. This concept is the primary principle behind the power of electromagnetic energy.

## Inherent Energy of EM Burst

Each burst of electro-magnetic energy has an inherent amount of energy. The inherent energy is contained in the strengths of the electric and magnetic fields.

Specifically, the amount of energy is seen in the thickness of the threads of the electric and magnetic fields.

Using our analogy of the hand-woven bag, each thread represents a line of energy. A thicker thread means that the line has more energy. A thinner thread means the line has less energy.

This can be represented a line of energy where the thickness of the line corresponds to the amount of energy in each line. You can also view this same diagram as a thread of yarn as in our analogy, where the thickness of the line corresponds to the thickness of the yarn.

This can also be represented by arrows, where the direction of the arrow indicates direction of flow, and thickness of arrows corresponds to the amount of energy in each line.

Figure 4.1

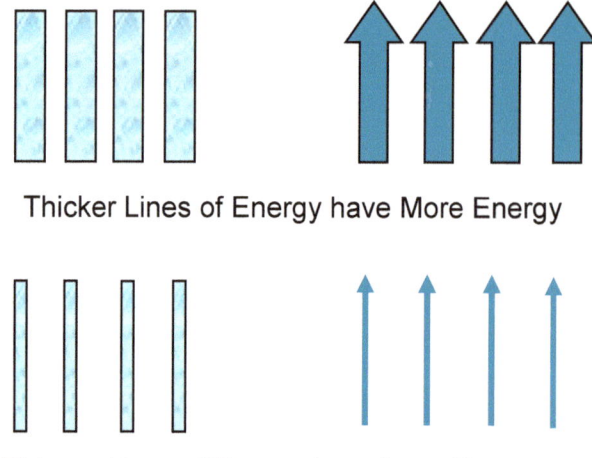

Thicker Lines of Energy have More Energy

Thinner Lines of Energy have Less Energy

## Kinetic Energy of EM Burst

Introduction

As stated above, each burst of electromagnetic energy has an inherent amount of energy. The inherent energy is contained in the strengths of the electric and magnetic fields. However, this energy can be observed in several different ways.

There are several ways in which the energy of the EM burst is exhibited: 1) forward motion, 2) frequency of pulsation, and 3) amplitude of pulsation. Each of these manifestation of energy are energy of motion, aka Kinetic Energy. The total combination of kinetic energies in our burst of EM energy is what we measure as the "energy value" of our individual burst.

Remember that kinetic energy is the energy of motion. (This is contrast to stored energy). Whenever an object moves, in any type of motion, that object is exhibiting kinetic energy. The concepts of forward motion, pulsation, and amplitude are each types of energy, specifically kinetic energy.

> The kinetic energy in EM Bursts is exhibited by forward motion, frequency of pulsation, and amplitude of pulsation.

If you want to increase the energy, the first thing to do is to choose a higher frequency of pulsation. After that if you want to increase the energy (yet keep the same frequency) then you add energy into the fields in order to increase the amplitude. Note that the speed of the forward motion cannot be changed, so you will not be able to adjust the energy in that area. Each of these statements will be explained below.

Electromagnetic Energy: The Composition of the Burst

Because we are talking about energy in detail, at this point we should clarify the differences between "electromagnetic energy" and "kinetic energy".

Electromagnetic energy is the type of energy the object is made of. Kinetic energy is how that energy exists moment to moment.

As an analogy: electromagnetic energy is like string wrapped in a ball. Kinetic energy is like throwing that ball of string through the air, or tugging on some individual strings.

Forward Motion

The first and most obvious energy of our EM burst is the energy of forward motion. Remember that our energy burst is created as a sphere, like a baseball. Like a baseball, this sphere of energy can be thrown through the air. When the baseball or energy sphere travels through the air, the object has forward motion – the kinetic energy of forward motion.

With a baseball, the amount of energy in the forward motion is provided by the baseball player. The energy of forward motion is exhibited as the speed in which the baseball travels. However, in the burst of electromagnetic energy things are slightly different. Like the baseball, the energy of forward motion is exhibited as the speed in which the burst of EM energy travels. However, the speed of forward motion is the same for all types of electromagnetic energy. This amount cannot be changed.

Therefore, the kinetic energy due to forward motion is exactly the same for all frequencies of electromagnetic energy.

> The speed of forward motion is the same for all types of electromagnetic energy.

> The kinetic energy due to forward motion is exactly the same for all frequencies of electromagnetic energy.

## Frequency of Pulsation

As stated earlier, pulsation is a type of motion, and this motion is a type of kinetic energy. Also as stated earlier, a faster pulsation means the EM burst has higher kinetic energy. Conversely, a slower pulsation means that the EM burst has lower kinetic energy.

Therefore the amount of inherent energy contained in a burst of EM energy is most commonly observed as the frequency of pulsation.

## Amplitude of Pulsation

One of the ways kinetic energy is exhibited in electromagnetic energy bursts is through the amplitude of pulsation.

Remember that the "amplitude" is the stretch in our pulsation cycle. In our analogies, this is the furthest we can stretch our balloon or hand-woven bag. In our EM burst the amplitude is the fullest distance the electromagnetic fields stretch, in either direction, during the pulsation cycle.

Anytime you stretch an object this requires energy. Imagine stretching a rubber band, a shirt, a balloon, or a hand-woven bag – any of these items require energy to stretch. The same can be said for our burst of electromagnetic energy. In order to stretch our EM burst we must apply energy. (Specifically, the EM fields within the burst push outward). Therefore the amplitude is indeed an observable effect of kinetic energy.

Furthermore, in order to stretch an object a greater distance we must apply more energy. You know this when you stretch a rubber band or shirt outward a further distance. The same is true for our burst of electromagnetic energy. In order to reach a further amplitude, a greater amount of energy must be applied. Therefore, the size of the amplitude is directly correlated with the amount of energy applied to the stretch of pulsation.

# Increasing Energy

Introduction

We can add energy to the individual bursts of EM energy. (Note that we are still talking about individual bursts here. For multiple bursts, see the later sections in this chapter).

In general, when we increase the energy of the EM burst we are increasing the energy of the electric field, increasing the energy of the magnetic field, or both.

What this means on a practical level will depend on how we add the energy. If we keep the frequency constant, then the additional energy will go into the amplitude. If we keep the amplitude constant, then the energy will go into the frequency.

And if we apply the energy to additional molecules or additional wires, then the energy will create multiple bursts instead of just one. (This is one way we create greater power – see sections below).

Also note that it is best to add energy to the EM energy sphere before it is emitted. It is much more difficult to change the energies of EM bursts after being emitted.

Forward Motion

The speed of forward motion is a constant for all frequencies of electromagnetic energy. Therefore, adding energy will do nothing to increase the forward motion.

Increasing Frequency and Decreasing Wavelength

One of the main results of adding energy to our EM energy is to increase the frequency. When we add energy to the EM energy fields, this will often cause the fields to pulsate faster. The energy fields will complete the pulsation cycle more quickly (which means a higher frequency). This in turn means a cycle will be completed in a shorter distance (lower wavelength).

Increasing the Size of the Amplitude

We can increase the Amplitude of our energy burst. This is one of the ways we can change the energy of the burst. We can also use changes I amplitude for variations in communication.

Note that this change in amplitude must be done before we emit the burst of electromagnetic radiation. Every burst of EM energy has a fixed amplitude after it is created. The EM burst will stretch out to this fixed distance but no further. Unless influenced by a different burst of EM energy, this amplitude will not change.

However, we can change the amplitude before we emit the EM burst. What we do is add more energy to the electromagnetic energy, yet keep the same frequency of pulsation. That additional energy will go into the amplitude, allowing the fields to stretch out to a greater distance.

More specifically, we add energy to our energy fields. This is like making a longer string. Therefore the energy fields can push outward a longer distance.

This is much like weaving a larger hand-woven bag. With the existing hand-woven bag, there is a limit to how far we can stretch the bag. The length of the threads will naturally limit how far we can stretch the bag. Therefore, if we want to stretch the bag further, we will need more material. When we sew together this slightly larger hand-woven bag, we can stretch it more.

So it is with the electromagnetic fields: if we want our fields to expand out more in the directions of pulsation (yet keep the same frequency and wavelength), then we must add more energy to the energy fields (those interwoven lines of energy). By adding more interwoven lines of energy, when we send the EM burst out into the world, those fields will pulsate out a slightly further distance. And this is also known as a greater amplitude.

<u>Size of Original Sphere</u>
Note that the size of the original sphere may change when we add energy. Consider the hand-woven bag analogy above. We added more material, which created a slightly larger bag. From that, we could stretch the bag out more.

Similarly, when we add energy to the electric and magnetic fields in order to change the amplitude, we may in fact be increasing the overall size of the energy sphere as well.

Therefore, when we add energy to the EM burst (before emitting), while keeping the frequency and wavelength constant, then that energy will go into both a) increasing the size of the original sphere, and b) increasing the amplitude (stretch of pulsation).

# Power of EM Energy: Group Power

On a practical level, one of the most desirable things to do with electromagnetic energy is to increase the power. There are several ways to increase the power, many of which will be discussed throughout this book. Yet there is one main method for increasing the power: emitting multiple bursts of energy at the same time.

So far we have been talking about one burst, with one frequency and one energy level. But suppose we emit many bursts of the same frequency at the same time? Not just one by one in a line (as a series of bullets), no this is much more than that. We are emitting many bursts of the same size, from the same general area, at the exact same time.

> "Power" in Electromagnetic Energy is the number of bursts of the same frequency, emitted from one location, which are sent at the same time.

Each burst is the same amount of energy, but because these many bursts are packed together in a group, the overall effect is that of one large burst as compared to many smaller bursts.

Figure 4.1  Single Bursts Emitted Sequentially, like a Particle Gun

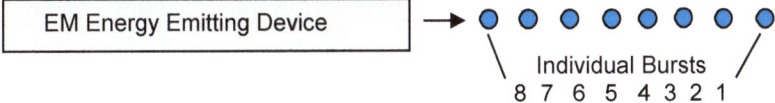

Figure 4.2: Multiple Bursts Emitted at the Same Time, Making a Pack of Bursts, and therefore Greater "Power" in each "emitted group"

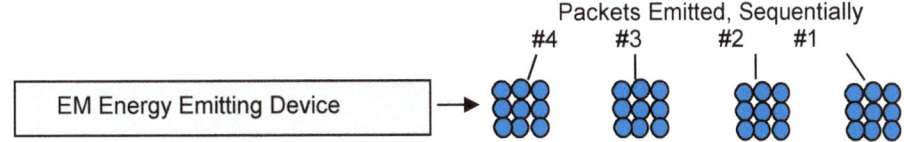

Packet of Bursts. Each Packet has more bursts, and therefore more power, than any single burst alone.

## Increasing the Power

Every time we emit another burst of EM energy at the same time as all the others, we increase the overall power. Therefore, one of the most effective ways to increase the power of electromagnetic energy is to emit more individual EM bursts at the same time.

> To increase the power of electromagnetic energy, we send more bursts of energy at the same time.

Figure 4.3: Increasing Power by Making Larger Packets of Bursts

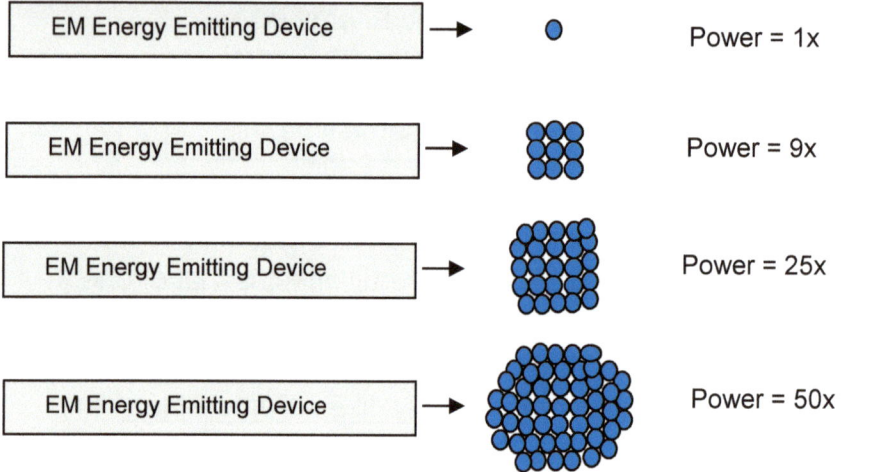

## Focus and Intensity

There is another way to increase the power of electromagnetic energy, and that is through using focus.

Focus can be considered two things. It can either be considered as aiming, or it can be considered as concentrating. Aiming will direct the electromagnetic energy where you desire. However, aiming it will not do anything to increase the power.

We want to think of focus as the act of concentrating. Using a focusing device, we can take separate paths of electromagnetic energy, then channel them into one small area. In essence, we are packing together numerous bursts of electromagnetic energy (you will recall that this is the definition of power of EM energy).

The most common focusing devices use lenses or mirrors to concentrate the bursts of EM energy.

A common focusing device is a magnifying glass. The electromagnetic energy from the sun comes down in many separate paths. Yet using the magnifying glass we can channel many of those paths into one small area. We are concentrating the electromagnetic energies, and therefore we are creating more power.

If you hold the magnifying glass to a section of grass, you can start a fire. This is because you have concentrated the energies so much onto that small area that the grass will burn.

Therefore, another one of the best ways to increase power is to use some type of focusing device. This will concentrate the paths of electromagnetic energy into a small area, and therefore increase the intensity (aka power) at a particular point.

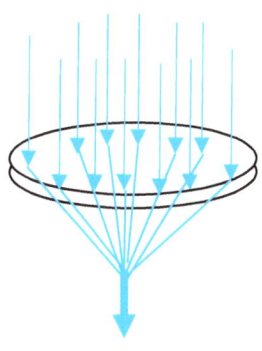

# Chapter 5
# Energy Spread of EM Energy

## Introduction
Electromagnetic energy will spread out as it travels forward. There are two ways this occurs: group spread, and individual burst spread.

The first mechanism refers to the package of bursts spreading apart as they travel. The second mechanism refers to individual bursts spreading out their electromagnetic fields.

## Group Spread of Energy Emission
The first method for energy spread to occur is through group spread. In this process, a grouping of EM bursts start as being bunched together, then gradually spread out as they travel forward.

Most emissions of electromagnetic energy do come as a single burst, but rather as many bursts emitted together at one time. Recall from our discussion on power of electromagnetic energy that the very definition of power is a package of individual energy bursts emitted from one location at the same time. Also recall that greater power is achieved by sending out a larger package of energy bursts at the same time.

Figure 5.1: Examples of Packages of EM Bursts
These are bursts of EM energy of the same frequency and amplitude, which emitted from the same location, at the same time.

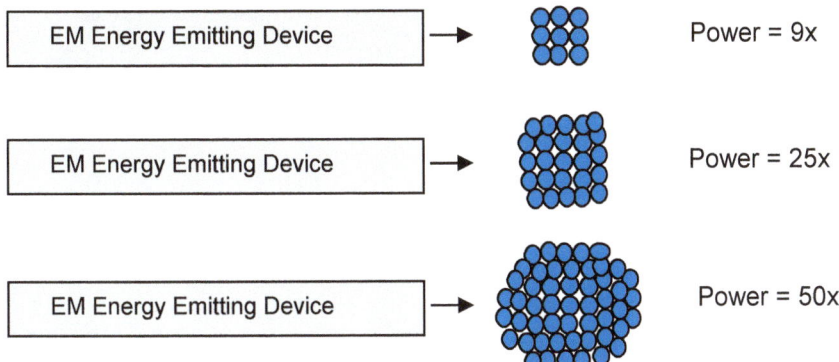

After a large package of energy burst is emitted, this package will travel forward and will spread out at the same time.

Remember that these bursts are individuals. It is not one large burst, but rather many individual bursts. Also remember that these bursts are not physically linked, and they are only weakly held together. Therefore, this package is simply a large group of individual bursts. As such, these bursts will spread out from each other, at a standard rate, as they move forward.

Figure 5.2 Example of Group Energy Spread

Individual bursts in initial emission package will separate as they travel forward.

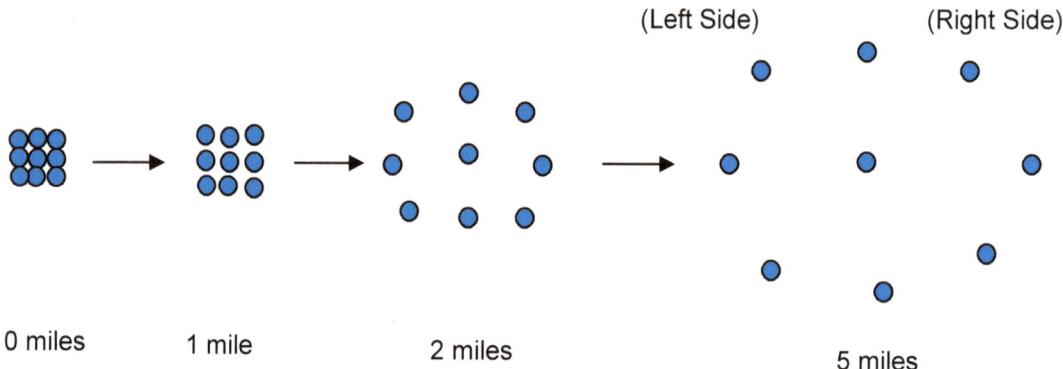

## Frequency, Wavelength, Amplitude with Group Spread

Notice that with the group spread the only thing that changes is the individual energy bursts separate. The frequency, wavelength, and amplitude remain the same.

Because the individual bursts of energy still exist – they are just further apart from their fellow bursts of energy – then all other factors for each energy burst remains the same.

Specifically, the frequency of pulsation for each burst remains the same. Consequently, the wavelength (distance to complete one pulsation cycle) also remains the same.

In addition the amplitude remains the same; the distance to which each burst pulsates in each direction does not change.

Therefore in group spread, the only thing that changes is that the individual energy balls separate. The individual balls of energy remain unchanged.

However, the power received will be lower. (See other sections for details).

> During group spread,
> the frequency, amplitude, and energy
> of each individual burst
> remains unchanged.

## Stars and Flashlight as Group Spread

The most common examples of Group Spread are the stars. Why are the stars so faint when compared to our Sun, especially when many of those stars are larger than our own? It is because of the distance and the Group Spread of the EM energy bursts.

If we were close to any star we would be overwhelmed with the size of the package of each emission. Even if there were only one small region of the star emitting energy, we would be overwhelmed, simply because of the number of EM energy bursts being emitted at one time from one location is enormous.

Yet coming from so far away, we only receive a handful of bursts. All the other bursts are spread thousands of miles apart, with a total spread of at least million miles. Therefore all we see of this original powerful emission is a few of the individual bursts.

Therefore the closer we are to the original object, we will receive more of the original EM energies. We know this as a brighter light, a greater intensity, a stronger signal, or other related terms. Further away, we receive only a handful of the original energy bursts, while the majority of bursts are spread widely over us and to each side. The result is a fainter signal, because that is all we receive.

This is just one example of the Group Energy Spread. Another example is the flashlight. The light starts as a focused point. Ahead of you, the light spreads out, in a uniform circle. You also notice that the light is strong close to you, but a few feet ahead the light dissipates, and there is a limit to how far you can see. This is another example of Group Energy Spread.

All frequencies EM energies will spread in this way with distance. Any frequency of EM energy emitted as a group of numerous bursts will start out strong, and spread out evenly over distance.

Note that the amount of power received as a result of group energy spread is an important practical factor when working with electromagnetic energy. These concepts will be discussed in the chapter on antennas.

# Spread of Individual Energy Burst

## Introduction

The second way in which electromagnetic energy spreads involves the individual energy burst. Specifically, the electromagnetic energy field, in each individual energy burst, will spread out as it travels.

Keep in mind that here we are looking at each individual EM energy burst. This is contrast to the above mechanism which involves the grouping of bursts. Also, remember that we are talking about the spreading of the electromagnetic field, whereas the above mechanism was the spreading of individual bursts.

## Individual Burst is Interwoven Lines of Energy Fields

As discussed in previous chapters, a burst of electromagnetic energy is made of criss-crossing lines of energy. These lines are the paths of electromagnetic energy and magnetic energy. The general appearance of a burst of EM energy is that of a hand-woven bag. In our case, the burst of energy has interwoven energy fields.

## Fields Changing Shape, and Spread of EM Energy Field

These fields can change shape. For example, the fields expand and shrink repeatedly. As discussed earlier, this is the pulsating action of the EM burst. This is what creates the frequency. Also as stated earlier, the amplitude is the distance which the fields expand in each direction.

In addition to this pulsating expansion and contraction, the energy fields also spread out. Each time the field pulsates, it grows just a bit bigger. However, it has the same mass, which means that to grow bigger the energy field spreads out, always wider, yet always a bit thinner. Always larger diameter, yet always more space between the lines of energy.

Figure 5.3: Example of Individual Energy Burst Spreading
This is the spread of the energy burst (spread of the EM Energy Fields) as the burst travels forward. Showing the contracted state (the smallest state in each pulse sequence)...at sample distances.

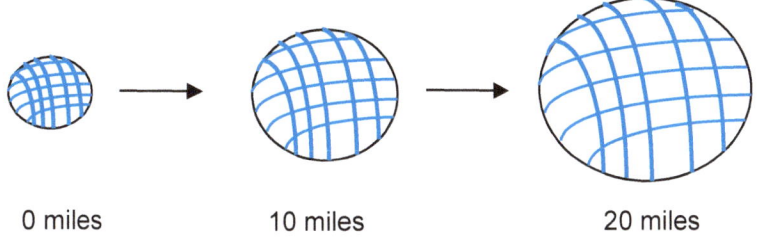

0 miles      10 miles      20 miles

## Spread of Individual Energy Burst versus Amplitude

### Introduction

At this point I want to emphasize that Amplitude is not necessarily changed with the energy spread.

Remember the difference between Amplitude and Diameter of Energy Burst. The amplitude is how far out the pulse stretches in each direction from its center point. However, the diameter of the energy burst is the distance from the center point to the edge, in the compacted state.

### Analogy of the Stretched Shirt

Here is an analogy to see how changing one does not necessarily change the other. Consider an average shirt. It has a basic size. You can stretch it a bit when you put it on and take it off, and this stretching will not impact the shirt. This is like the Amplitude.

However, you can stretch a shirt "out of shape", where it will no longer fit the same way. Areas of the shirt are now stretched larger, and they will never again back to the original shape. The spreading of electromagnetic field is very similar, where the field changes to a new larger shape, and will never go back to a smaller size.

And yet, there are areas of the shirt even in the stretched state that you can pull on just a little, such as the sleeves, and those areas will go back to their original state (to their new original state). The Amplitude is very much like this. Even in the new stretched state of the energy burst, the energy burst will pulsate out as it always has, and return to its "new compacted size".

## Options for Amplitude and Energy Spread

As stated above, the Amplitude is not necessarily affected by the energy spread. Most often the amplitude will remain the same regardless of diameter of the energy burst.

For example, if the Amplitude is 5 centimeters then that will be the amplitude of the pulse regardless of the compacted size of the burst. The diameter of the compacted burst could be a millimeter, a meter, or a kilometer, and the amplitude will still remain 5 centimeters.

On the other hand, the amplitude may change at some point. The electromagnetic field is a shape-shifting ball of energy. Therefore, while it is spreading out, it may also change the distance of pulsation. This is a possibility.

## Figure 5.4 Example of Energy Spread, AND showing that Amplitude may be the same in all.

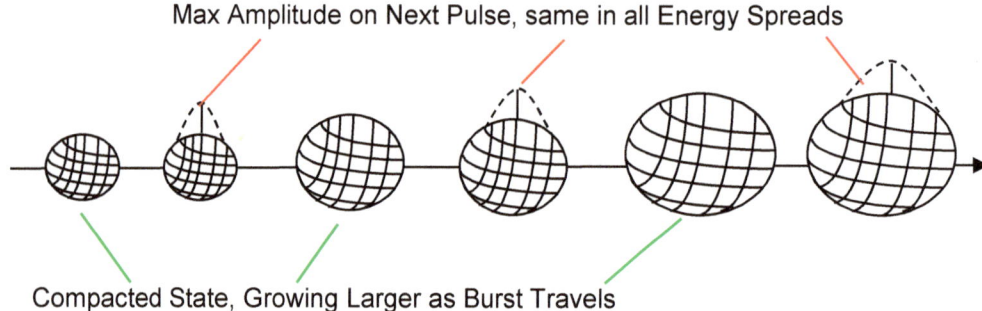

## Amplitude and Energy Spread vs. Amplitude and Original Energy

At this point I would like to emphasize the difference between the amplitude change with energy spread, versus the amplitude change with a larger initial burst of energy.

The most accurate way to change the amplitude of an EM burst is to change the amplitude before we send it out. Specifically, we add to the length of the energy strings.

This is done for each burst, before it is emitted. Then all bursts emitted at the same time (from all the molecules or electrons in the region) will have the same amplitude.

After the EM burst has been emitted, the amplitude will generally stay the same. The EM burst can travel millions of miles with the same amplitude.

However, because the EM burst contains strings, those strings can rearrange during travel. This is particularly true if acted upon by a force – even from afar – such as gravity, or the bending of space-time.

# Chapter 6
# Energy Strings and General Structure of EM Bursts

## Introduction

If you could hold a Burst of Electromagnetic energy in your hands, what would you see? If you could study a photon closely, if you could turn it in your hands, what would you discover? In this chapter I provide those answers.

In this chapter we will look at the fundamental nature of the EM Burst. What is it? What does it look like?

Also note that this chapter has many new concepts and models on the fundamental nature of EM bursts. These models will explain a great number of properties and processes related to electromagnetic energy.

## Energy Segments

It is time to get a more advanced understanding of electric and magnetic fields. The lines of the electric and magnetic fields are not simply single lines. Rather, each energy line actually consists of energy segments.

Figure 6.1: Showing Energy Segments in Each Energy Line

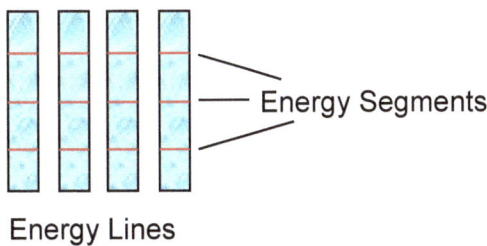

Energy Lines

These energy segments can separate. They can become more sparse, or can remain more compact. This an important concept when discussing the size of the EM Burst.

> Each line of energy consists of individual energy segments.
>
> These segments can exist close together or spaced very widely apart.

## Figure 6.2

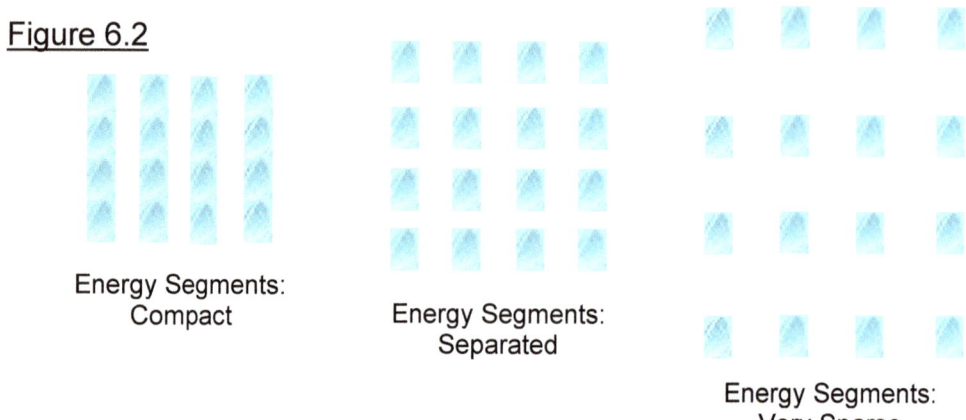

Energy Segments: Compact

Energy Segments: Separated

Energy Segments: Very Sparse

## Figure 6.3

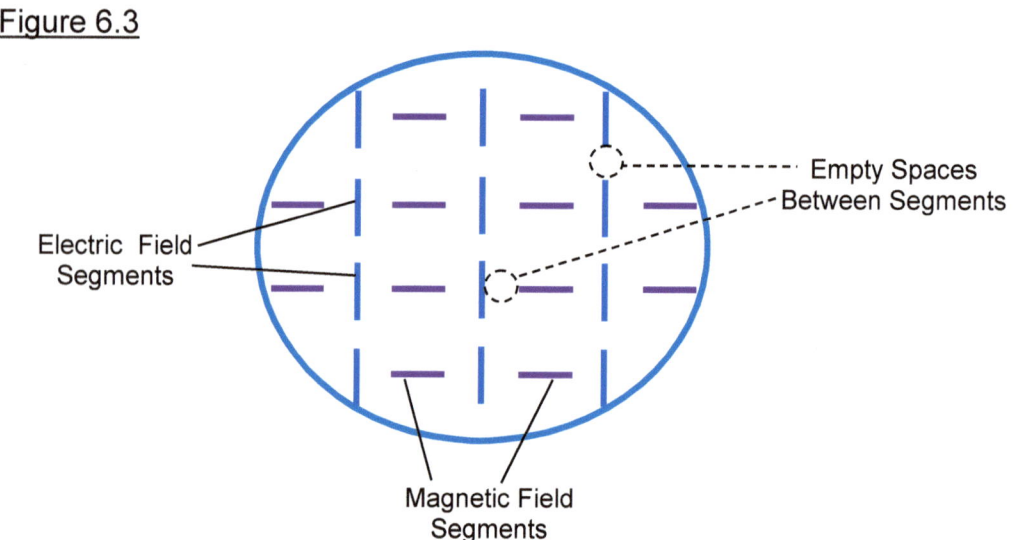

Electric Field Segments

Empty Spaces Between Segments

Magnetic Field Segments

# Energy Strings in a Burst of Electromagnetic Energy

## Introduction

One of the most important concepts to understand in relation to electromagnetic energy are the "Energy Strings". These energy strings are the cause for almost everything about electromagnetic energy, including frequency, wavelength, energy, emission, absorption, cause of pulsation, and much more.

## The Energy String

The energy string is a special entity in the universe. It has both mass and energy, as dual components of its basic nature.

Einstein proved long ago that all mass has energy, and all energy has mass. This holds true for every type of object, and every type of energy, though it is not always easy to see. However, with the energy string, mass-energy relationship is very fundamental. The energy string is essentially the "missing link" between mass and energy.

Thus, the energy string is an object. This object has certain dimensions, including thickness, length, and mass. Yet at the same time this object is a container of energy. This physical object consists almost entirely of energy.

---

Energy Strings:

1. Energy Strings are physical objects
    with both Mass and Energy.

2. The Energy String is an object with dimensions
    including thickness, length, and mass
    yet the string is a container composed of energy.

3. Energy Strings are the entities in the universe
    which provide the direct link between mass and energy.

---

## Similar to Virus (inanimate versus living being)

The energy string is analogous in nature to the virus. In biology, the virus is the link between animate (living beings) and inanimate (non-living objects).

The virus can stay dormant for years. Yet given the proper conditions, the virus becomes a living entity. The virus feeds, reproduces, and evolves. Thus, the virus is a biological link between the animate world and the inanimate world.

Similarly, in the area of physics the energy string is the link between the world of mass and the world of energy. The energy string is a very basic entity (much like the virus is a basic entity) which is simple in nature, yet bridges two different worlds.

## New Meaning of Energy Strings

From now on we will use the term "energy strings", and this term will mean any single entity of energy. What we previously called an energy line or an energy segment are now referred to as energy strings. The particular energy string may have different thickness or length, but the term "energy string" will in fact refer to any independent entity of energy, of any size, in our burst of electromagnetic energy.

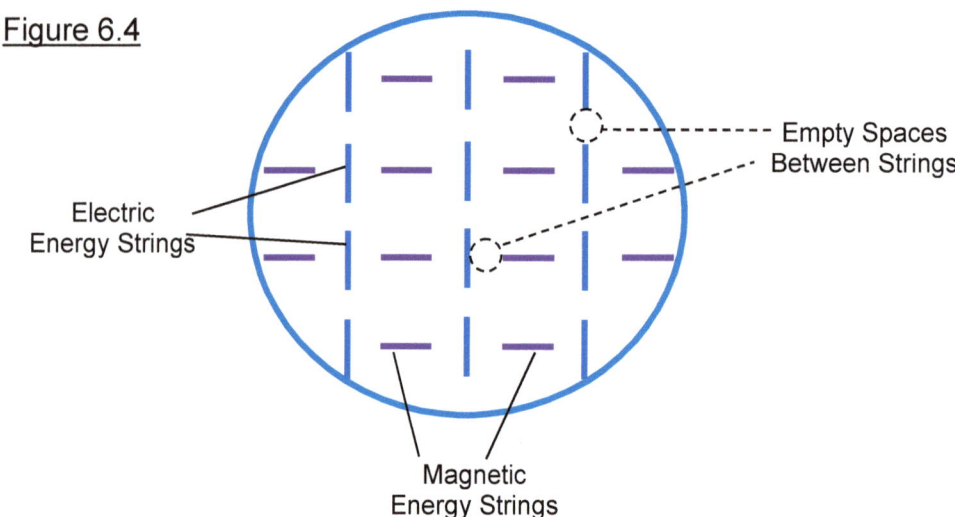

Figure 6.4

### Specific Details of Energy Strings

The concepts of energy strings are of such importance for our understanding of electromagnetic energy (as well as electrons) that we must understand exactly what these energy strings are.

Much of these concepts will be discussed in this chapter. Many more concepts will be discussed (and illustrated) throughout the book. Here are a few highlights:

Energy strings have both mass and energy. When energy strings are created, they are created with a certain amount of mass, and a certain amount of energy.

Each energy string is an independent object. Each string can move around in various ways. Yet energy strings can join and separate.

The total number of energy strings, the total mass of energy strings, the total energy of the strings, as well as how the strings are arranged, will determine all the properties of electromagnetic energy.

## Compare and Contrast to General String Theory

### Introduction

At this point we should make some comparisons to general string theory as commonly described by physicists. This section is mostly for the readers who are familiar with "string theory" commonly used by physicists, and to show how my understanding of strings in the photon is very much different.

If you are unfamiliar with string theory, just skip this section, as the general string theory does not apply to anything discussed in this book. Stay with my descriptions of energy strings, and you will understand every concept of electromagnetic energy, and every discovery explained in this book.

### Energy Strings in Photon are NOT the strings of "String Theory"

In brief: the energy strings which compose the electromagnetic energy are NOT the same as the energy strings commonly described by physicists. There are similarities, but it is important to know that there are differences.

Therefore, as you read my book you must stay with my definition and understanding of energy strings.

## For This Book: Energy Fields are Strings, Everything Else is Particle

"String Theory" is broad term. Ask several physicists what "string theory" means, and you will get several answers. A common description is that particles are not exactly particles, but are in fact strings. In contrast, I am not ready to say that all particles are strings. However, I AM CONFIDENT that what we know as energy fields are in fact energy strings.

## For This Book: Strings Move but do NOT Vibrate

Another aspect of string theory I have heard is that the vibration of strings in the string theory create the entity. For example, lower vibrations of energy strings make the entity appear as an object, higher vibrations of energy strings will appear as dark matter, higher vibrations yet will appear to us as dark energy, and the highest vibrations will appear as pure energy.

While I like some concepts of this theory, this is really very much separate from the energy strings in the photon. The general string theory is similar to our photon strings in that the same strings provide both mass and energy.

I am also intrigued by the idea that the vibration of the energy strings is the mechanism which deems the object primarily mass or primarily energy. However, I have not investigated these concepts thoroughly. More important, those concepts of string vibration are not relevant to understanding anything discussed in this book.

Thus, for this book know that energy strings do have both mass and energy. Yet the exact relationship between mass and energy within the energy string are not important.

Also, for this book know that energy strings are objects which can move. They migrate, they push, and they pull. The objects join and they separate. However, do not think about any vibration. It is very important that you understand that the "wave" from our electromagnetic energy, as well as the "wave" from our photon, comes from the collective motions of energy strings, and NOT from the individual vibrations of energy strings.

### Restating the Highlights of Energy Strings

Our new understanding of energy strings is different from the energy strings discussed in general string theory. Yet there are similarities. For this book, and all the new discoveries presented in this book, we must focus on the understanding of energy strings as discussed here. Set aside any previous concepts of energy strings you may have learned elsewhere.

As stated above: in our photon, energy strings have both mass and energy. When energy strings are created, they are created with a certain amount of mass and a certain amount of energy. Each energy string is an independent object. Each string can move around in various ways. Yet energy strings can join and separate.

The total number of energy strings, the total mass of energy strings, the total energy of the strings, as well as how the strings are arranged, will determine all the properties of any particular burst of electromagnetic energy.

## Thickness and Length of Energy Strings

### Overview: Energy String Factors and Resulting Total Energy

The dimensions of the energy strings will determine the total energy in the EM burst. There are three main factors:
1. Thickness of the Energy Strings
2. Length of Energy Strings
3. Total Number of Energy Strings

When we increase any of these factors (thickness, length, or total amount of energy strings) then the overall energy of the EM burst will increase.

### Thickness, Length, Amount and the Exhibited Effects

Although each energy string factor will contribute to the overall energy of the EM burst, each factor will do so in different ways.

1. Thickness: The thickness of the energy strings will determine the frequency of the EM burst. (Details will be discussed in subsequent chapters).

2. Length: The length of the energy strings will determine the Amplitude of the EM burst. (Details will be discussed in subsequent chapters).

3. Total Number of Energy Strings: The total number of energy strings, in conjunction with the thickness of the energy strings, will determine the final frequency of the EM burst. (Details will be discussed in subsequent chapters).

Frequency, Amplitude, and Energy Strings

It is worthwhile to restate the above in a different way:

a. If we want to increase the frequency of the EM burst, we can either add more strings to the system, or increase the thickness of the energy strings.

b. If we want to increase the amplitude of the EM burst, it is easiest to add to the length of the energy strings.

---

Thickness and Length of Energy Strings:

1. Thickness of energy strings determines the frequency of the electromagnetic energy:
   - Thicker energy strings will create faster frequency.
   - Thinner energy strings will create slower frequency.

2. Length of energy strings contributes to the amplitude:
   - Longer energy strings will result in greater amplitude.
   - Shorter energy strings will result in smaller amplitude.

# Mass of Energy Strings

## Introduction

Energy strings have mass. This is a very important concept to remember, because it is will be a central point in the cause of EM pulsation.

## Thickness and Mass of Energy Strings

A thicker energy string not only has more energy, but it also has more mass.

Energy strings are unique entities in the universe. They are a hybrid of energy and mass. These energy strings have characteristics of both energy and particles. As such, each energy string has a certain amount of mass. Therefore thicker strings have more mass, and thinner strings have less mass.

Note that to some degree all solid particles and all types of energy are a hybrid of energy and mass. It is just some entities exhibit more mass, and other entities exhibit more energy. Energy strings are one of the few entities where both characteristics are almost equivalent.

This is also an important concept, which will be explored in detail later.

## Number of Strings, Total Energy, and Total Mass

A similar concept is the number of energy strings in our EM burst. In every EM burst, we have numerous energy strings. As discussed earlier, these strings can separate or recombine throughout the life of the EM burst. However, there is a total energy, and a total mass.

Imagine that all strings were combined into one sold entity, just one string. This string would have a total amount of energy. This string would also have a total amount of mass.

Again, these concepts of total mass and total energy in our EM burst will be important later. We will discuss the details in subsequent chapters.

# Separating, Traveling, and Joining of Strings

## Introduction

The energy strings within the EM burst are able to separate, travel, and join together. This forms the basis of the pulsation of the EM burst, as well as the amplitude.

## Separating and Joining Within the EM Burst

Each energy string is able to separate from the other string. Segments of energy strings can be pulled off and become separate entities. These become independent energy strings unto themselves.

These strings can also combine together. Energy strings can be pulled together, combine, and thereby make a larger energy string. (This larger string can be either additional thickness or length). Therefore we have a dynamic situation with these energy strings. These strings can join, separate, and rejoin…in any combination, and at any time.

More details will be described in subsequent chapters. For now it is important to know that the movement of energy strings, the separation of energy strings, and the rejoining of energy strings, together play important roles in the frequency of the EM burst.

# Snapshot of the EM Burst

## Introduction

What does the EM Burst look like? If you could hold it in your hand and study it, what would you see?

If you take a snapshot of an EM burst at any one time, you will see numerous energy strings, inside a basic volume of space. These energy strings exist perpendicular to each other, and there are spaces between each energy string. They have varying thickness and length.

## Basic Sphere of the EM Burst

Note that because the EM burst is dynamic, at this point it simplest to look at the EM Burst when it is the basic sphere. This sphere is how it exists before it pulsates in either direction.

The number of strings, the thickness, and the space between each string (in this basic sphere stage) are factors which are all determined at the time of the creation of the EM burst. This basic sphere does not change much over time. Each time the EM burst returns to the basic sphere shape, its dimensions will remain essentially the same.

For example, below we have two EM bursts. The first one is a high energy, high frequency, small diameter sphere. The second is a lower energy, lower frequency, larger diameter sphere. These spheres were created that way at the time of emission, and every time either sphere returns to the stage of basic sphere shape, the dimensions will remain the same as before.

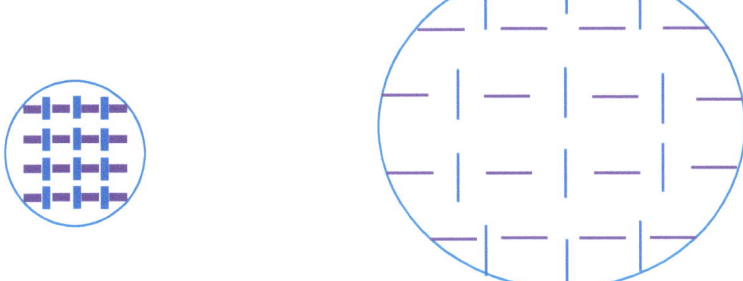

Note that detailed explanations of diameter of sphere, as well as related factors such as frequency and energy, will be explained in subsequent chapters.

# EM Core

There is another factor in the basic structure of the EM burst: the EM Core. Some EM bursts, particularly the highest frequency EM bursts, will have a Core.

This core is a compact region of energy strings. In this core the energy strings are tangled and knotted up together. They are a dense mass of energy strings, which will never separate and never migrate.

This core is only created for the highest energy EM bursts. And it is the density of this core which in fact creates the high frequencies of these EM bursts.

Details of the EM core will be explained in subsequent chapters.

# Energy Strings in Emission and Absorption

## Introduction

Now that we have discussed the basic structure of the EM Burst, let us discuss a related concept: how energy strings provide the process of emission and absorption.

Since we are discussing the basic concepts of energy strings within an EM burst, it is also a good time to discuss the processes of energy strings as related to the EM burst and other objects.

As stated above the energy strings within the EM burst are able to separate, travel, and join together. These properties also are the basis of the emission and absorption of EM energies.

Details will be described in subsequent chapters, but for now we can view some basics.

## Energy Strings When Emitted

Briefly, whenever an EM Burst is created, it is being "built" with energy strings. Every EM burst which is emitted is in fact created with a certain number of energy strings, with specified thicknesses and lengths.

These properties, when the EM Burst is created, will determine how the EM Burst pulsates and how it interacts with various objects.

Again, details will be discussed in subsequent chapters. This is just a brief overview. The important things to know for now are that 1) energy strings are put into the EM burst before it is emitted, and 2) once the EM burst is emitted the properties of the energy strings will determine everything about how the EM burst acts.

## Energy Strings Upon Absorption

Briefly, we know that when that EM burst can be absorbed. Specifically what occurs is that energy strings are leaving the EM burst and becoming attached to the other object.

These energy strings will remain contained within the EM burst as long as the EM burst travels freely through space. Yet, when the EM burst interacts with another object, such as an electron or another EM burst, then energy will be transferred. More specifically, some energy strings will be transferred.

Similarly, we know that EM bursts often deflect, and change to a lower frequency of color. The actual process involves transfer of energy strings. Some energy strings go to the new object. This leaves the EM burst with fewer strings, which is less overall energy, and therefore a lower frequency.

Again, details of all this will be explained in subsequent chapters. For now it is enough to know that when energy is exchanged between an EM burst and another object, that energy transfer occurs through the movement of energy strings.

# Summary of All Concepts Discussed in this Book, Focusing on New Discoveries and Models

Overview
　　The following are ALL the major concepts presented in this book. New theories and new models which are created by the author will be highlighted with the following: (new)
　　More specifically, ALL concepts and models discussed in this book will be listed here, with notes as I believe to be important. All concepts, models, and analogies will be listed by chapter.
　　Yet, the vast majority of concepts presented in this book are new concepts. The majority of concepts are in fact new discoveries. Each new discovery will be highlighted.

Purposes of this List
　　There are several purposes of this list:

　　1. The reader can easily see, at a glance, each concept. The reader can refer to this list and see the main concepts presented throughout the book. This I provide as another type of teaching aid.
　　This is the main purpose of the Summary of All Concepts. It is an aid for the reader, to refer to these pages, to read and re-read the summaries, in order to get a better understanding of all concepts presented in this book.

　　2. There are many new discoveries presented in this book. Therefore the reader can refer to this list. Because many concepts are new to the world they are only presented here, and therefore should be understood as easily as possible. Again, as an educational aid, the reader can see the summary points of all concepts.

　　3. Furthermore, the reader can see how these many new concepts are interrelated. As the reader flips through the pages and reads various concepts, he can see how these concepts are interrelated. This will give the reader a much broader and much more subtle understanding of everything discussed in the book.

Also, by flipping through the shorter summaries the reader can see how the new concepts and new models explain multiple processes and multiple observed phenomenon. Overall, the reader will gain a much better sense of how these new concepts must be true...because they all fit together in such a tidy and comprehensive way.

4. Of course I want personal credit for all new discoveries. I also want credit for new models, new theories, new analogies, and new illustrations. Therefore I want to emphasize to each reader which of those discoveries and models are mine.

Certainly I am proud of all of my discoveries, all of my models, and all of my creations. And certainly I want to receive proper credit for these. Therefore, by highlighting those new discoveries and models, readers (particularly those in the scientific community) can appreciate what I have to offer, and will acknowledge me for my contributions.

## Organization of the Summary List

These summary points are organized chapter by chapter. Then, for chapters which are longer or which have more information, the summary points are further subdivided by general topic.

When the summary point is a new concept, the word "new" will be listed after the summary point.

## Significance of my Discoveries, and Credit for my Contributions

In addition, many of these new concepts will be further elaborated on, in terms of what makes the concept new, or why the concept is a significant advancement in our understanding of science.

These additional elaborations are mostly written for readers who do not have a detailed background in quantum science, so that they can appreciate the significance of the discoveries.

Also, some of the concepts will seem so "obvious" when explained that it will be too easy for others to take the credit for my work, or to say "I already knew that" when in fact they did not.

Therefore, while it is true I mostly want to present these discoveries for the benefit of society, I also want proper credit. I am simply elaborating with enough details to ensure that my contributions are indeed credited to me.

# Chapter 1: Understanding Electromagnetic Energy
## Summary of Concepts

1. Electromagnetic Energy is a pulsating ball of energy which travels through space (partially new)
   *Newton and Einstein knew that electromagnetic energy is a particle. Feynman also believed that it is a particle. Yet there is still much debate generally.
   * I am the first to understand that this particle pulsates as it travels through space.

2. Each signal of electromagnetic energy is a distinct burst of pulsating energy, with its own properties, and traveling in its own way through space.

3. Electromagnetic energy is NOT a wave. Electromagnetic energy is a particle, not a wave. However, the burst of electromagnetic energy can be tracked as a wave. (new)
   *This is a very important concept. For example, Mars is a planet, Mars is not an ellipse. Yet the path of Mars can be tracked as an ellipse. In the same way, electromagnetic energy is a sphere, not a wave, yet the path of this sphere can be tracked as wave.
   This is a fundamentally different way of looking at electromagnetic energy. It changes your perspective dramatically. And it is totally my idea.

4. Electromagnetic energy has the following basic properties or motions:
   a. It is a distinct object like a baseball
   b. It pulsates like a heart, or inflating a balloon (new)
   c. Its path can be tracked as a wave (new)
   d. Groups of bursts spread out like pizza dough (partially new)

5. A burst of Electromagnetic Energy is like a Baseball. It can be emitted or thrown into the air, it will fly through the air in a straight line, until it is caught, absorbed, or bounced back.

6. A burst of Electromagnetic Energy pulsates like a Heartbeat. Each pulse looks like a balloon inflating and deflating, where the "balloon" is filled with electromagnetic energy. (new)

*I am the first person to discover that the EM burst has a pulsation, like the heartbeat or the balloon. This understanding leads to understanding many further concepts of EM energy.

7. The rate of the pulsation is commonly known as the frequency. (new)

*The concept of frequency has long been known and studied. However, I am the first person to present exactly what frequency is: the frequency is the rate of pulsation of the EM burst.

8. Electromagnetic Energy is not a wave. The motions of the electromagnetic energy can be tracked as a wave, but the object itself is a ball of energy, not a wave. (new)

*As said above, this is very different way of looking at electromagnetic energy. Looking at electromagnetic energy as a particle which can be tracked as a wave, rather than a mysterious "wave" entity, makes much more sense, and leads to many more discoveries.

9. Electromagnetic Energy is like Pizza Dough. Emission begins as a big ball of grouped EM bursts, then spreads out in a perfect circle, getting thinner with each pulse, as the individual photons spread further apart.

(Partially new)

*It has long been known that electromagnetic energy spreads out as it travels, and this can be measured by the inverse square law. However, most scientists are not clear (even if they know this themselves) that emission starts as a group of bursts and then spreads out.

Also, the pizza dough analogy is mine. Hence, we begin with the group of EM bursts like the ball of pizza dough, then the bursts spread out...exactly like pizza dough spreading out - in an even circle. Also the dough continues to spread out wider and wider. Soon holes appear between the dough, just as black spaces appear between the EM bursts. Eventually the ball of dough becomes very wide – many feet, many miles – and the holes are larger than the bits of dough between the holes, which is the same as when the bursts spread so wide that the individual bursts reaching our eyes are barely noticeable.

# Chapter 2: Pulsating Energy Fields
# Summary of Concepts

## General New Concept of the EM Burst

10. A burst of electromagnetic energy is made of criss-crossing lines of energy. A burst of electromagnetic energy looks very much like a hand-woven bag. (New)
    *A fundamental new discovery of electromagnetic energy is that it looks very much like a hand-woven bag, where the lines of yarn in our photon are electric and magnetic energy strings.

11. Like a hand-woven bag, the burst of electromagnetic energy can stretch and change shape. (New)

## Details of Energy Strings and Energy Fields

12. The thickness of the energy strings, the length of the energy strings, and the total amount of energy strings will determine the properties of the electromagnetic energy burst (including dimensions and flexibility), much in the same way that in our hand-woven bag the thickness, length, and total amount of yarn will determine its dimensions and flexibility. (New)

13. What is commonly referred to as an energy field is in fact a set of energy strings in motion. (New)
    *The concepts of electrical energy fields and magnetic energy fields have been known about for a long time. However, I am the first to understand that these fields are actually energy strings.
    The energy strings are essential to understanding numerous processes involved in magnetism, electrical current, and electromagnetic energy.

14. Energy strings are physical objects which have both mass and energy. The energy string has thickness, length, and mass, yet it is essentially a container of pure energy. (New)
    *This is a big concept.

15. There are two types of energy strings: electrical energy strings, and magnetic energy strings. (Partially New)
   *It is well known that there are two energy fields. However, I am the first to restate this as "there are two types of energy *strings*". This is more than just semantics; it is a different conception of the physical reality.

16. Energy strings flow in a particular direction. (Partially New)
   *It is well known that energy *fields* flow in a particular direction – particularly from a power line or from a magnet. However, I am the first to state that the energy fields are actually a set of strings flowing in a particular direction.

17. Energy strings can move about, depending on the circumstances. (New)
   *I know that energy strings will not only flow, but they can move about. In electrons most energy strings are attached to the electron, yet some energy strings remain free. In the burst of electromagnetic energy the energy strings now only flow in alternating directions, but can migrate as well. I am the first person to understand all of these type of energy string motions.

18. In various aspects of nature, electrical energy strings and magnetic energy strings can exist as separate entities. (Partially New)
   *It is important to emphasize that electrical energy strings and magnetic energy strings are different entities. And they can be different amounts in different materials. For example, a given material may have a set of strong magnetic energy strings, yet a set of weak electrical energy strings.

19. When we have a burst of electromagnetic energy, the burst contains both electrical energy strings and magnetic energy strings. (New)
   *Maxwell first determined that electromagnetic energy consists of both electric and magnetic fields. However, my understanding of electromagnetic energy is totally different from anything proposed previously. Thus, I put forth the statement that electromagnetic energy is primarily a burst, not a wave, and that this burst is comprised of electric and magnetic energy strings. This is a totally new concept.

20. In a burst of electromagnetic energy, the electrical energy strings and the magnetic energy strings always flow perpendicular to each other. (Partially New)

   *It is well known that energy *fields* are perpendicular to each other, yet I state this for the energy *strings*.

   Also, I am the first to propose that because the electrical and magnetic energy strings will flow perpendicular to each other in an electrical current, which will create the perpendicular arrangement of strings in the EM burst.

   I also am the first to demonstrate in a subsequent chapter why the two types of energy strings will always be perpendicular.

## New Concept of the Photon

21. When the burst of electromagnetic energy is initially created, it appears much like a ball of yarn, with criss-crossing segments of electrical energy strings and magnetic energy strings. (New)

   *This is a totally new model, and it is the first model in the book to represent what a photon actually looks like.

## The Pulsation Process

22. The Pulsation of the Electric and Magnetic Fields is like two perpendicular balloons inflating and deflating, or two hand-woven bags stretching and compacting.

   *This is a totally new understanding. This visual image will help our understanding of many electromagnetic energy processes.

23. As the energy fields pulsate outward, they extend primarily along two vectors, each perpendicular. For example, the electric energy field may extend upward, while the magnetic field extends to the left. (This is very well known).

24. The strength of each energy field is primarily along a straight line vector, yet the field also extends outward in a balloon like shape.
   *Partially new concept

25. The Pulsing of the Electric and Magnetic Fields Outward is like two perpendicular balloons inflating or two perpendicular hand-woven bags stretching.

*As mentioned above, this is a new concept.

26. The pulsation motion of the electromagnetic energy actually consists of two fields pulsating outward, inward, then outward in the opposite direction.

For example, the pulsation may be the following set of steps:
a. An electric field stretches upward and the magnetic field stretches to the left.
b. Both fields retreat.
c. The electric field stretches downward and the magnetic field stretches to the right.

The net result is a pulsation motion, similar to a heartbeat, with two perpendicular pulsations outward, retreating, and then two pulsations in the opposite directions from before. (New)

*This is a totally new model.

## One Merged Region to Simply our Two Field Regions

27. There is also a region in space where the two fields overlap. This region of space can be considered the "electromagnetic field", a single entity. This is contrast to the "electric field" and the "magnetic field" which are two different entities.

Thus, this region of space, this "electromagnetic field", consists of electric energy strings and magnetic energy strings, in the same region of space. (New)

Using this single "electromagnetic field" can simplify our drawings and understanding of processes related to electromagnetic energy.

*This is a totally new model.

28. We can draw this single "electromagnetic field" only as the merged region of electric strings and magnetic strings, while at the same time not including the rest of the electric field or the rest of the magnetic field.

The net result is volume of space, again shaped like a balloon, which is "filled" with electrical strings and magnetic strings.

The location and direction of this balloon shape is in between the two balloon shapes of the electrical pulse and the magnetic pulse. For example, if the electric field pulsates upward and the magnetic field pulsates to the left, then this single "electromagnetic field" pulsates to the upper left (followed by retreating, and then stretching to the lower right).

(New)
*This is totally new to the world.

29. In total, the pulsation of electromagnetic energy, looking only at the single merged region of energy fields, consists of the following set of steps:
   a. Energies Begin as Compact Sphere
   b. Combined Energy Fields Extend One Direction
   c. Energies Retreat to Compact Sphere
   d. Combined Energy Fields Extend Opposite Direction
   e. Energies Retreat to Compact Sphere

   This process repeats essentially forever.
   (New)
   *This is a totally new model.

### Frequency of Pulsation

30. The frequency of a burst of electromagnetic energy is simply the rate at which the pulse goes through one complete cycle. More specifically, the frequency of an electromagnetic energy burst is the time it takes for the electrical field and the magnetic field to complete one full cycle of pulsation in all directions.
   (New)
   *New understanding, based on new models of pulsation.

31. Both the electrical field and the magnetic field pulsate at the exact same rates. This is due to the inherent nature of the cause of pulsation (as discussed in later chapters). Therefore, the burst of electromagnetic energy burst will have one frequency, never two.

32. The frequency of electromagnetic energy is measured in cycles per second. This means the number of times the electric and magnetic fields complete full pulsation cycles in each second.

# Chapter 3: Creation of Wave Patterns and Particle-Wave Duality
## Summary of Concepts

### Creation of the Wave Pattern

33. A burst of electromagnetic energy is a sphere of energy. Its motion can be tracked as a wave, but physically it is a sphere not a wave. (new)
    *This is a totally new concept, and a major new understanding to the world.

34. The creation of the wave pattern is a combination of the forward motion of the energy burst, and the repeated pulsing of the electromagnetic fields. (new)
    *This is a totally new concept, and a major new understanding to the world. The diagrams in chapter three show exactly how this occurs.

35. The Wave Pattern is created by the combination of the pulsating sequence and the forward motion of the energy sphere.
    Specifically, the electromagnetic energy pulsates in opposite directions, similar to inflating and deflating balloons, while at the same time the burst flies through the air in a straight line. When you plot these motions together, you can see the wave pattern. (new)
    *This is a totally new concept. This is a MAJOR insight into electromagnetic energy. This concept explains the particle-wave duality which scientists have baffled scientists for over 100 years!

### Forward Trajectory and Frequency

36. The forward trajectory of all bursts of electromagnetic energy will be at the same speed in a particular medium. Most often, we look at electromagnetic energy as it travels through space or air. The speed of the forward trajectory in space and air is constant. (The air will absorb or deflect the photon, but not significantly alter its forward speed).

37. The frequency of the EM burst will depend on the internal properties of the photon (as will be discussed later). Thus the frequency of pulsation of the EM burst will vary from photon to photon, while all the forward speed of all photons will be the same.

## New Understanding of Wavelength

38. The Wavelength of the burst of electromagnetic energy is the forward distance required for the EM burst to make one complete cycle.
    (Partially new)
    *This is a new understanding, using a combination of traditional understanding of waves, and my new understanding of pulsation and wave patterns.

39. Remember that the wave *pattern* is created as a combination of the pulsation and the forward trajectory. Therefore it makes sense that the *particular value of the wavelength* is a combination of the particular pulsation frequency and the particular forward speed.
    For example, a faster pulsation means that the photon travels only a short distance before completing its pulsation cycle. Therefore the faster pulsation results in a shorter wavelength.
    Conversely, a slower pulsation means that the photon travels a much longer distance before completing its pulsation cycle. Therefore the slower pulsation results in a much longer wavelength. (New)
    *This is a very important concept. Understanding the exact relationship between pulsation frequency and wavelength provides us with a much deeper understanding of electromagnetic energy.

40. Because the forward speed is constant for all EM bursts in the same material, then as we compare EM bursts in the same medium (such as air) the wavelength is primarily dependent on the pulsation frequency.
    Thus, we most often make comparisons between EM burst in the same medium – such as air, space, or water. Because we are talking about the same medium, the forward speed of all EM bursts are the same. Therefore we can neglect the particular speed for our discussions, or use the same value of the particular speed in our calculations.
    However, if we were to compare the same energy photon in different mediums – such as traveling through water versus traveling through air, then we must take into account the forward speed (as well as other processes).
    This concept is generally beyond the scope of this book, however we should remember that throughout this book when we neglect to discuss the speed of the photon, we do so only because we are talking about all photons as traveling through air.

# Energy as the Primary Property of EM Burst

41. The primary quantity of electromagnetic energy is in fact ENERGY. It is not frequency, wavelength, or amplitude. Rather, all of those other properties are in fact related to the primary essence of the photon: that of its inherent energy. (Partially new)

   *This concept isn't really discussed. Although the very name of electromagnetic energy is "energy" the energy concept of the EM burst is often secondary in discussions, while frequency and wavelength are discussed first. Thus I want to emphasize the concept that Energy is the central property, and all other aspects come from that property.

42. Every burst of electromagnetic energy is created with an inherent amount of energy. The energy is contained within the Energy Strings. The total amount of energy contained in all of the energy strings is the total amount of energy in the photon.
   (New)
   *This is a totally new concept to the world

43. "Motion" is in fact a way that an entity exhibits energy. If the object moves faster, it does so because it has more inherent energy. If the object moves slower, it does so because it has less inherent energy.
   This concept is true for all objects, including photons. Thus, a photon which pulsates faster does so because it has more inherent energy. Conversely, a photon which pulsates slower does so because it has less inherent energy.
   (Partially new)
   *The concepts of motion and internal energy have long been understood. I am adding the concept of how this relates to photons.

44. The total energy of the photon, as contained in the specific energy strings, will determine the motions of the photon. This is primarily the rate of pulsation, but can also be seen as additional forward speed, and other more subtle motions. (Partially new)

45. The frequency of pulsation of the EM burst is caused by the internal energy. In other words, it is the internal energy which causes the pulsation frequency, it is NOT the frequency which creates the energy of the photon.
(New)
*This is another very important concept. Most scientists discuss frequency and energy by saying "the photon has X frequency, and therefore has Y energy". I know that the cause and effect are in fact reversed. It is the internal energy which causes the frequency; it is not the frequency which causes the energy. Understanding the proper cause and effect for pulsation frequency will help us understand the properties of electromagnetic energy much more accurately.

### New Understanding of Amplitude

46. The Amplitude of the electromagnetic energy burst is how far out the electric and magnetic fields stretch during each pulsation. Using our analogy of the hand-woven bag, the amplitude is the maximum distance to which we can stretch our hand-woven bags. (New)
*This is new, based on my new understanding of electromagnetic energy as interwoven energy strings, and the process of pulsation.

### Particle-Duality Solved and Explained

47. Particle-Wave Duality Summarized: The concepts of why electromagnetic energy is both a particle and a wave are extremely important. These concepts solve a mystery which the greatest scientists have been puzzling over for over 100 years. Now I have solved it. Because of the significance, it is important to summarize the concepts:

   a. Particle: A burst of electromagnetic energy is foremost a particle. It is a sphere of electric and magnetic energy fields. This sphere of energy fields flies forward through space similar to a baseball.

   b. Wave: The burst of electromagnetic energy is not a wave. Rather, the motion of the burst can be tracked as a wave. The wave motion of the EM burst is created as a combination of two motions: the pulsation motion and the forward motion.

c. The EM burst pulsates as it travels. Each energy field pulsates outward in one direction, then collapses inward, and then pulsates outward in the opposite direction. This pulsation repeats continuously. At the same time, the EM burst is flying through the air. This looks very much like a baseball as it flies across a field. The pulsation motion and the forward motion exist together, at all times. When you look at these combined motions from the side, and trace the path, you will indeed see a wave.

d. Particle-Wave Duality: Therefore, again, the Particle-Wave Duality for electromagnetic energy can be explained as follows: any burst of EM energy is primarily a particle, yet the combined motions of the particle (the pulsation and forward trajectory) can be tracked as a wave.

# Chapter 4: Energy and Power of EM Energy
# Summary of Concepts

## Inherent Energy and Total Energy of All Strings

48. Each burst of electromagnetic energy has an inherent amount of energy. This inherent energy is the total amount of energy contained in all of the energy strings. (New)

49. The total energy of the EM burst will be seen as motions. More specifically, the total energy of the energy strings will cause the motions of the EM burst. These motions include: frequency of pulsation, amplitude of pulsation, and forward motion. (New)

50. The total inherent energy of the energy strings can also be said to be the kinetic energy of the photon. Remember that kinetic energy is the energy of motion, and therefore the kinetic energy of the photon is the amount of motion observed. This includes the specific frequency, the specific amplitude, and the specific forward speed. (Partially New)

51. If you want to increase the energy, the first thing to do is to choose a higher frequency of pulsation. After that if you want to increase the energy (yet keep the same frequency) then you increase the amplitude. Note that the speed of the forward motion cannot be changed, so you will not be able to adjust the energy in that area. (New)

52. The way to increase the overall energy to a photon is to add energy to the energy strings. Specifically, you can add energy in three ways:
    a. Add energy to the thickness of the existing energy strings
    b. Add energy to the length of the existing energy strings
    c. Add more energy strings, of any thickness or length.

  *This is a totally new concept, and one of my major insights. These insights explain much to us regarding the fundamental entities which operate in the universe.

### Energy Amounts of Individual Energy Strings

53. Each individual energy *string* has a specific amount of energy. The energy of the particular energy string depends on the thickness of the string and the length of the string. (New)
    *This is a totally new concept, and one of my major insights to the world.

54. Thicker energy strings contain more energy. This is very similar to yarn: Just as a thicker piece of yarn has more material, a thicker energy string has more energy. (New)
    *This is a totally new concept, and a major insights to the world.

55. Longer energy strings also contain more energy. Just as a longer piece of yarn has more material, a longer energy string has more energy.
    (New)
    *This is a totally new concept, and a major insights to the world.

56. Thicker energy strings will result in faster pulsation frequency. This is due to the inherent nature of the cause of pulsation. These concepts are more fully developed in other chapters.
    (New)
    *These two concepts are HUGE new insights to the world.

57. Longer energy strings will allow a greater amplitude. This is similar to longer yarn in a hand-woven bag allowing us to stretch the bag out further.
    (New)
    *This is also a major new concept

58. Adding additional strings will also increase the pulsation frequency. Again, this is due to the inherent nature of the cause of pulsation.
    (New)
    *This is also a major new concept

59. Remember that thicker energy strings will increase the frequency of pulsation, while longer energy strings will increase the amplitude. This is an important clarification to remember.

(New)
*These two concepts are HUGE new insights to the world. These insights are major. The difference between thickness and length of the strings, and how each produces a different effect is a major new understanding of EM properties.

### Individual EM Bursts and Groupings of EM Bursts

60. Electromagnetic energy can be emitted in a series of bursts, like a particle gun.

61. However, more often electromagnetic energy is emitted as group of bursts. This group of bursts are photons of identical frequency, emitted from the same general location at the same time. Historically, such a group of bursts has been called a "packet" or a "quanta" of photons.

The region from which the bursts are emitted is usually small, such as a square inch. Each singular burst is emitted from an individual molecule in the material or from an electron on a wire. Because it is the same material, the same frequency of EM will be emitted. Thus, it is common for dozens of identical EM burst to be emitted from the same small region of material at the same time. The net result is a "group" of identical EM bursts emitted from the same location at the same time.

(Partially New)
*My clarification helps the reader understand the concept of group bursts or packets – what are they, and why they are emitted as such.

Also, certainly the idea of "quanta" or "packets" was applied to electromagnetic energy early on. But did those scientists understand the concepts of group bursts of electromagnetic energy as precisely and accurately as the way I describe? Perhaps yes for some scientists, perhaps no for other scientists. That is why I state my concept is "partially new".

## Power in Electromagnetic Energy

62. "Power" in Electromagnetic Energy is the number of bursts of the same frequency, emitted from one location, which are sent at the same time. For example, a cluster of 24 identical photons has greater power than a cluster of 9 identical photons. (Partially New)

"Intensity" is another term for Power of electromagnetic energy.

*I believe I am the first person to express Power of Electromagnetic Energy in this precise way.

Furthermore, there may be other engineers who intuitively understand the concept, but then again probably many do not. Thus, the concept may be totally new for some (which makes it a "new" concept), or for others this concept may be vaguely understood yet more accurately clarified (thus a "partially new").

63. One of the ways to increase the power of electromagnetic energy is to increase the number of identical bursts emitted at the same time. For example, if we increase the number of identical bursts emitted at one time from 10 to 30, we will have increased our power by a factor of three.

(Partially New)

*Radio engineers know how to increase the power, yet may not understand the actual process of increasing the cluster of photons. Thus this concept is "partially new"

64. Another way to increase the power of electromagnetic energy is to use a focusing device (such as a lens or a mirror).

The focusing device works by diverting all the arriving photons to a central location. This will concentrate the paths of electromagnetic energy into a small area, and therefore increase the intensity (aka power) at a particular point.

This focusing device is primarily used at the receiver, but can be used prior to transmission as well.

(This concept is well known. However I do offer my clarification and explanation).

# Chapter 5: Energy Spread of EM Energy
## Summary of Concepts

65. Electromagnetic energy will spread out as it travels forward. This is due to group spread.

After a large package of energy burst is emitted, this package will travel forward and will spread out at the same time. This spread occurs in an even circular pattern, much like the spreading of pizza dough.

66. The intensity of electromagnetic energy will decrease as the electromagnetic energy travels forward. This is because the individual bursts are spreading out, and the number of bursts reaching your eye or any detector will become fewer and fewer as the detector is further from the source.

This is the process which creates the intensity of the stars. Closer stars such as the sun appear very bright because most of the clusters of photon bursts are still close together. In contrast, stars billions of miles away appear only as faint dots because the individual bursts have spread very wide apart (and we only see a few clusters of photons).

67. In group spread the only property that changes is that the individual photons spread further apart. All other factors remain the same, including the forward speed, frequency, wavelength, and amplitude.

# Chapter 6:
# Energy Strings and General Structure of EM Bursts
# Summary of Concepts

## Details of Energy Strings

68. Energy strings are fundamental to understanding of all properties of electromagnetic energy. A detailed understanding of energy strings include the following:

    a. Energy Strings are physical objects with both Mass and Energy.

    b. The Energy String is an object with dimensions including thickness, length, and mass.

    c. The Energy String is essentially a container composed of energy.

    d. Energy Strings are special entities. Energy strings provide the direct link between mass and energy.

(Very New)
*There were others who came up with the term energy string. However, my views of energy strings differ greatly from the common scientific vernacular of "energy strings". Therefore, this is all completely new, all are my contributions, and very important.

*These Energy Strings are fundamental to many processes and particles in our universe. Knowing about these Energy Strings, understanding their properties, and appreciating how they operate, will explain many phenomenon of our world.

*Therefore, the entire concept of these strings, and how they operate, is a major contribution to the world. I am the first to propose these strings and their properties.

69. Because the energy string has both energy and mass as the primary components, the two are directly related. Thus, when the energy string has more energy, the string also has greater mass. And when the energy string has less energy, it will have less mass. (new)
    *These concepts of Energy and Mass in one object make many important properties. Einstein was the first to say Energy has mass and mass has energy. Yet I am the first to state this as a direct connection in energy strings, and to state the properties which derive from this concept.

70. Similarly, there is a correlation between the amount of material, energy, and mass. For example, when an energy string is thicker or longer, the energy string will have more energy, and it will have greater mass. Conversely, when an energy string is thinner or shorter, the energy string will have less energy, and it will have less mass.
(new)
*These correlations are totally my idea, my presentation to the world. And these correlations will have many important effects.

71. What is commonly known as an "energy field" is in fact a set of parallel energy strings. (New)
*This is totally my concept.

72. Furthermore, each parallel energy string is actually not one single string, but rather is a set of energy string segments.
Thus, every "energy field" actually consists of numerous individual segments of small energy strings.
(New)
*This is totally my concept.

73. Energy strings can perform many actions, including:
a. Energy strings can migrate
b. Energy strings can join together
c. Energy strings can separate

(New)
*This is totally my concept.

74. In a burst of electromagnetic energy there are two types of energy fields: the electric energy field and the magnetic energy field. We now know that energy fields are in fact composed of numerous small energy strings. Therefore, in a photon we really have two sets of energy strings: electric energy strings and magnetic energy strings.
(New)
*The electric fields and magnetic fields have long been known about (introduced by Faraday). Yet I am the first person to state that these fields are in fact energy strings. This understanding is important to better understand the processes of electromagnetic energy.

# Details of the Photon:
## Totally New Understanding

75. When you hold a photon in your hands to study it, this is what you will see:

   a. The photon is essentially a sphere of energy.

   b. The photon contains a set of electrical energy strings and a set of magnetic energy strings.

   c. Energy strings lay parallel and perpendicular to other energy strings, with spaces between each energy string.

   d. Specifically, electrical energy strings lay parallel to each other, and magnetic energy strings lay parallel to each other. Then all electrical energy strings lay perpendicular to all magnetic energy strings.

   e. From afar, the photon looks similar to a ball of yarn. Zooming in closer, the photon has the appearance of a loosely woven bag with large spaces.

   f. A very close inspection of the photon reveals that the strings are in fact floating in air. Thus the strings are able to migrate and move, yet are also loosely held together by a type of gravitational pull.

   g. The size of the photon, the frequency of the photon, and the amplitude of the photon are all determined by the overall energy contained within all the energy strings of that photon. (This is discussed in later chapters).

   (New)
   *What a Photon Looks like if you held it in your hand…this is a big deal. I am the first to propose this entire construction of the interior of the photon. Thus everything I write here is very new to the world, and a very big contribution to science.

76. All energy strings are allowed to move, including migrating outward, retreating inward, and sometimes traveling in unusual directions.
   (New)

77. Small energy strings can join other energy strings, thus making a longer or thicker energy strings. Similarly, longer or thicker energy strings can break apart into smaller energy strings.
   (New)

78. They dynamic nature of the energy strings - how they migrate, how they travel, how they join and disjoin – result in many interesting properties of electromagnetic energy.
   (New)

79. Many burst of electromagnetic energy, particularly those with the highest energy (and greatest frequency) will have an "Electromagnetic Energy Core". This EM core is a compact region of energy strings where the strings are tangled together. The EM core is a dense mass of energy strings, which will never separate and never migrate.
   It is this density of tangled energy strings which causes the greatest internal energies and greatest frequencies of electromagnetic energy.
   (New)
   *This is totally my idea. This another addition to our model which does explain the fastest frequencies. And of course has the highest energies.

## Overview of Emission and Absorption of EM Bursts

80. The basic processes of EM emission and absorption are related to the energy strings.

In brief, emission involves the launching of energy strings from an electron. In brief, the absorption process involves the transfer of energy strings to an electron or molecule. These processes are discussed in great detail in later chapters. (New)

*This is very significant! I am the guy who first really understands what the process of emission of EM burst really means (which is the launching of energy strings). I am also the guy who first understood what absorption really means (the transfer of energy strings from the EM burst to the electron, atom or molecule).

Thus, this is a major, major discovery. And all details associated with this concept are important discoveries.

81. When a burst of EM is created, the EM burst is "built". Thus, the amount of energy strings, the thickness of the strings, and the lengths of those strings, in addition to an energy core (if there is one) are all "inserted" into the photon at the time the photon is created. These processes are discussed in great detail in later chapters.

(New)

82. After the EM burst is created and emitted, the EM burst becomes an independent entity. It travels through the air on its own, and pulsates on its own (according to the particular construction of the EM burst).

(New)

83. Full or partial absorption of a burst of electromagnetic energy is in fact full or partial transfer of energy strings from the photon to the absorbing electron or atom. These processes are discussed in great detail in later chapters.

(New)

www.ingramcontent.com/pod-product-compliance
Lightning Source LLC
Chambersburg PA
CBHW050727180526
45159CB00003B/1152